DEVELOPMENTS IN REINFORCED PLASTICS—5

Processing and Fabrication

CONTENTS OF VOLUMES 2–4

DEVELOPMENTS IN REINFORCED PLASTICS—5

Processing and Fabrication

Edited by

G. PRITCHARD

Department of Industrial, Organic and Polymer Chemistry, Kingston Polytechnic, Kingston upon Thames, Surrey, UK

ELSEVIER APPLIED SCIENCE PUBLISHERS
LONDON and NEW YORK

ELSEVIER APPLIED SCIENCE PUBLISHERS LTD
Crown House, Linton Road, Barking, Essex IG11 8JU, England

Sole Distributor in the USA and Canada
ELSEVIER SCIENCE PUBLISHING CO., INC.
52 Vanderbilt Avenue, New York, NY 10017, USA

WITH 30 TABLES AND 157 ILLUSTRATIONS

Softcover reprint of the hardcover 1st edition 1986

British Library Cataloguing in Publication Data

Developments in reinforced plastics.
5: Processing and fabrication
1. Reinforced plastics—Periodicals
I. Pritchard, G.
668.4'94'05 TP1177

The Library of Congress has cataloged this work as follows:
Developments in reinforced plastics.—1- —London; New York, NY: Applied Science Publishers, 1980–
v.: ill.; 23 cm.—(Developments series)
ISSN 0260-9185 = Developments in reinforced plastics.

1. Reinforced plastics—Collected works. I. Series.
TP1177.D48 668.4'94'05—dc19 83-640868

ISBN-13: 978-94-010-8356-0 e-ISBN-13: 978-94-009-4179-3

DOI: 10.1007/978-94-009-4179-3

The selection and presentation of material and the opinions expressed in this publication are the sole responsibility of the authors concerned.

PREFACE

Any series with a title beginning *Developments in* . . . is obviously intended to report innovatory and novel ideas. The trouble with innovatory thinking is that it often seems too esoteric for practical people to bother with. Certainly, this book is not meant primarily to be a quick-reference manual for fabricators. Its purpose is rather to signal the kind of developments which almost certainly will impinge on the world of reinforced plastics in, say, four or five years' time.

In this particular volume most of the authors have directly or indirectly addressed the practical problems of processing and fabrication with reinforced plastics. There has been no attempt to review the current state-of-the-art of producing fabricated articles in reinforced plastics by such techniques as filament winding or pultrusion because these subjects have already been well covered elsewhere. Nor have I even tried to provide a comprehensive survey of all that could be called new in this field. Instead, I have simply taken a number of important and somewhat underestimated topics, generally material-orientated rather than machine-centred, and asked leading figures to summarise the scene. At the risk of appearing arbitrary let us consider the first chapter by Cattanach and Cogswell. They tell us how a new material has been produced which not only adds to the range of composites available, it makes possible new fabrication processes (at least, new to FRP). Consequently it should result ultimately in many new markets and products. The opportunities are lucidly and imaginatively set out.

The other authors show in different ways how science is helping to monitor and control the processing of matrix resins, mainly crosslinkable polymers. The importance of reliable instrumentation for this purpose is stressed, and the rheology of the materials is given prominence.

After all, it is ease of fabrication rather than material costs which remains *the* reason for the commercial success of unreinforced thermoplastics. One day, the composites people will achieve the more difficult task of mastering the appropriate processing technology for their materials too, and those who follow their progress closely will reap corresponding benefits. The efforts of those researchers who contributed to this book will contribute to the achievement of this goal.

G. PRITCHARD

CONTENTS

LIST OF CONTRIBUTORS

A. Apicella

Polymer Engineering Laboratory, Department of Material and Production Engineering, University of Naples, Piazzale Tecchio, 80125 Naples, Italy

D. P. Bashford

Fulmer Research Laboratories Ltd, Hollybush Hill, Stoke Poges, Slough, Berkshire SL2 4QD, UK

J. B. Cattanach

Imperial Chemical Industries plc, Petrochemicals and Plastics Division, PO Box 90, Wilton, Middlesbrough, Cleveland TS6 8JE, UK

F. N. Cogswell

Imperial Chemical Industries plc, Petrochemicals and Plastics Division, PO Box 90, Wilton, Middlesbrough, Cleveland TS6 8JE, UK

Tsuneo Hirai

Faculty of Engineering, Department of Mechanical Engineering, Doshisha University, Imadegawa-Karasuma, Kamigyo-ku, Kyoto 602, Japan

David E. Kranbuehl

Department of Chemistry, College of William and Mary, Williamsburg, Virginia 23185, USA

D. Landman

Consultant, 515 Ramona Court # 20, Monterey, California 93940, USA

Charles Y-C. Lee

Materials Laboratory, Air Force Wright Aeronautical Laboratories, MLBP, Wright-Patterson Air Force Base, Dayton, Ohio 45433, USA

Horst Rothe

Institute for Technology of Polymers, Department for GRP, Hohe Strasse 6, 8010 Dresden, German Democratic Republic

Gottfried Wiedemann

Institute for Technology of Polymers, Department for GRP, Hohe Strasse 6, 8010 Dresden, German Democratic Republic

Chapter 1

PROCESSING WITH AROMATIC POLYMER COMPOSITES

J. B. CATTANACH and F. N. COGSWELL

ICI plc, Petrochemicals and Plastics Division, Wilton, UK

SUMMARY

Continuous fibre reinforced thermoplastics have only become commercially available in the form of preimpregnated tape since 1982. Most of the excitement associated with this new class of materials has been the combination of toughness and environmental resistance which makes them good candidates for high performance structural composite applications. However, by their very nature, thermoplastics offer new scope for high rate, efficient fabrication technology, the development of which may be seen as a key step in enabling high performance fibre reinforced plastics to become a mass production industry. In this chapter we outline the processing strategies which are currently being researched and provide background information on the associated sciences of rheology, heat transfer and morphology on which those strategies depend.

1.1. INTRODUCTION

Continuous collimated high strength carbon fibre preimpregnated with polyether etherketone (PEEK), a thermoplastic resin, is available from ICI under the generic title Aromatic Polymer Composite[1] which is usually abbreviated to the letters APC. This is the first commercial member of a new class of high performance composite materials based on thermoplastic resin systems, whose preparation and properties have been discussed by McMahon in Volume 4 of this series of publications.[2] The general characteristics of such composite materials have been reviewed[3] and a more specific mechanical property profile for APC can now be compiled from a range of independent published accounts.[4–8]

Thermosetting resins, and in particular epoxies, have, during the last decade, provided the major source of matrix materials for high performance structural composites based on continuous fibre reinforcement. Those resins have demonstrated a good balance of properties combined with convenient processing technology: the high performance of components made from such resins has generated a new design philosophy and a potential demand for an appropriate mass production process. As with any new high performance material the design philosophy has pushed the material to its limits and identified certain property constraints—shelf-life, water sensitivity and brittleness—which appear to be inherent in that family. Looking forward to the next generation of materials thermoplastics resins offer an attractive balance of properties to extend the use of composite materials.

The advantages offered by thermoplastic resins fall into four groups. The rigorous chemistry of linear chain polymerisation gives composite materials indefinite shelf-life from which the user obtains enhanced quality assurance. Toughness of the resin phase can translate into improved damage tolerance in the product, and crystallinity of the resin phase gives potential for excellent environmental resistance. Together toughness and crystallinity offer the product designer increased freedom to construct the best product. Finally, the thermoplastic characteristic of the resin permits: rapid processing, the ability to reclaim scrap via injection moulding, and new strategies to repair damaged structures. These advantages may be viewed as decisive factors in determining the future growth of the fibre reinforced plastics industry. That growth will be a partnership between the different classes of resin system, thermoset and thermoplastic each doing what it does best.

The recognition of the potential advantages offered by thermoplastic systems has led to the introduction of a number of preimpregnated tape products from a variety of sources including:

Aromatic Polymer Composites (APC) from ICI[1]
Ryton composites from Phillips Petroleum[9]
Torlon composites from Amoco[10]

Spiflex tape from SPIE Battignolles[11]
K Polymer composites from Dupont[12]
Fibre Imprégnée Thermoplastique (FIT) from Atochem[13]

Two distinct families are represented in this list. The polyimide systems—Torlon and K Polymer—are reported to have high glass transition temperatures, above 200°C, and are claimed to be tough in comparison with thermosetting resins. However, the recommended processing technology appears to include a post-cure process to obtain optimum properties. As linear chain, as opposed to crosslinked, thermosetting resins they may be considered to occupy a middle ground between crosslinkable epoxy and bismaleimide systems and true thermoplastics. The second group which includes Aromatic Polymer Composite, Ryton, Spiflex and Fibre Imprégnée Thermoplastique are based on polyether etherketone (T_g 143°C), polyphenylene sulphide (T_g 90°C) and nylon (T_g 60°C), respectively: all may be regarded as true thermoplastics in that no chemical change is invoked at the processing stage; all are semi-crystalline polymers having good solvent resistance.

All of these systems are described as thermoplastic composites, they all share some common features. Our experience is only with Aromatic Polymer Composite and before extrapolating the teaching of this chapter to other thermoplastic systems, including other systems based on carbon fibre and PEEK resin, the reader is recommended to study the data sheets for that system. Thermoplastic composite is potentially a very large field.

1.2. PREPREG FORM

APC-2 is supplied as a preimpregnated tape containing continuous collimated high strength carbon fibre. The tape is usually 140 mm wide. The standard product contains 60% by volume of carbon fibre having an areal weight of carbon of 145 g/m^2 giving a nominal ply thickness of 0·125 mm. As a uniaxial product this tape may be slit to form tapes of narrower width as required for particular processing applications.

The industry is continuing to research new product forms. One variant is the impregnation of single tow carbon fibre to give narrow tape of precisely controlled thickness and width. Such tape is especially suitable for weaving into broad goods. Another variant of

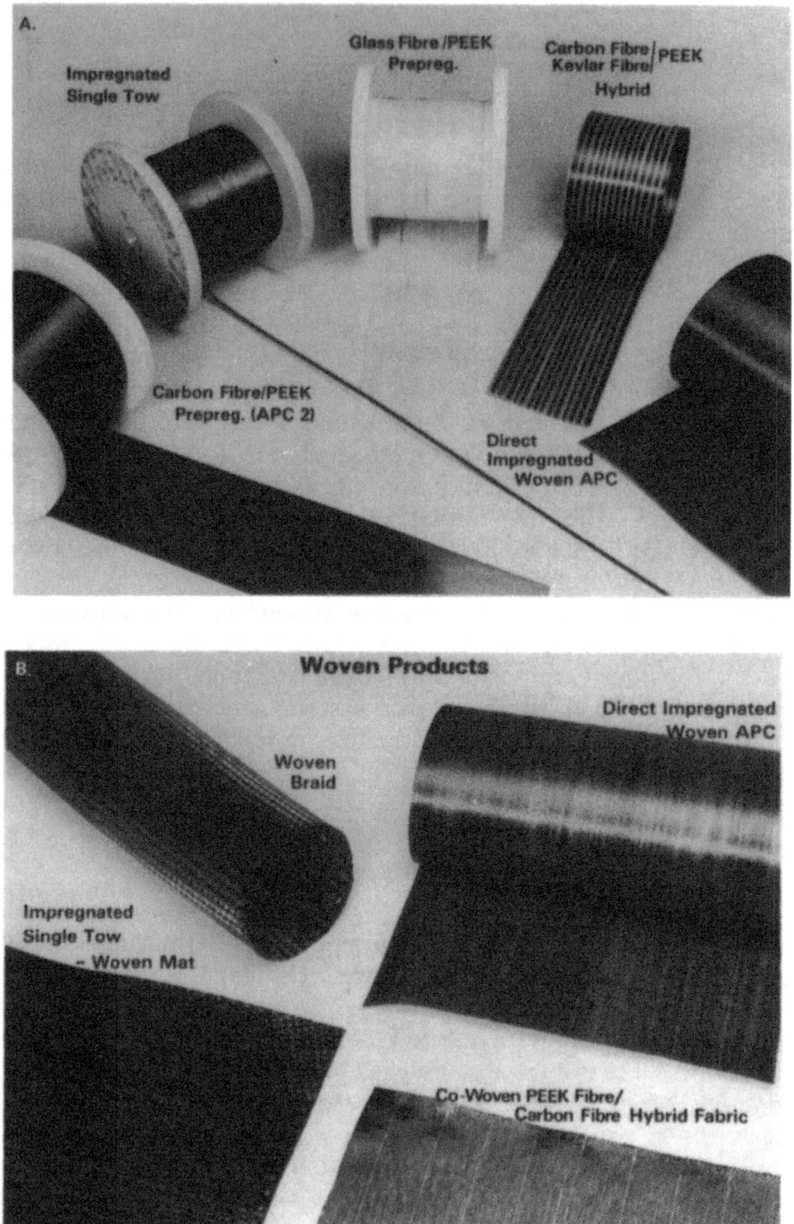

FIG. 1.1. (A) Preimpregnated tapes, (B) APC in fabric form.

simple prepreg form is crinkle cut tape which is fabricated directly from the basic prepreg by a heated crimping process.[14] This product can be exploited in certain fabrication processes requiring deep draw. Directly impregnated woven fabric is a further product form currently under development.

All APC systems are based on high strength carbon fibre and semi-crystalline polyether etherketone resin. Preimpregnated tapes based on glass fibre have also been produced as has prepreg using amorphous polyethersulphone Victrex resin and the general teachings of this chapter can be readily transcribed for those systems. Figure 1.1 displays a range of products developed so far. Of these, at the time of writing, only APC-2, based on continuous collimated fibre, is commercially available and in free supply.

1.3. HANDLING AND STORAGE OF APC

First a caveat: APC is stiff, thin and anisotropic. Rough handling, especially of the edge of the prepreg can produce splinters which will puncture the skin. Such splinters can be painful and the prepreg should be handled with care.

By comparison with prepregs of thermosetting resins, which have the properties of tack and drape at room temperature, APC prepregs are solid, although flexible because of their thinness. APC becomes tacky and can be draped at above its melting point, 343°C, and, for hand lay-up operations prior to a consolidation or fabrication stage, a heavy duty soldering iron (240 V, 175 W) is a valuable tool for tacking the prepreg together.

In APC thermoplastic prepreg the resin is fully polymerised and during any normal storage that chemistry does not change: APC does not require refrigerated storage. While it is recommended that APC should be stored in a clean, dry environment, if it should become contaminated it can be cleansed with appropriate agents without causing any deterioration to the product. Surface moisture should be dried off before fabrication with APC prepreg. APC-2 will pick up about 0·25% by weight of water, but this has no obvious effect on processing or properties. (Please note that this would not necessarily be true of other thermoplastic prepregs, in particular for those based on resins dependent on condensation polymerisation such as polyamides and polyimides). Considering the very high standards of housekeeping evident in laboratories handling high performance composites for aerospace applications there should be no problems encountered with APC systems.

1.4. CHARACTERISATION OF APC FOR PROCESSING

1.4.1. Density and Thermal Expansion

Polyether etherketone is a semi-crystalline resin whose density varies with temperature. Carbon fibres have an extremely low thermal expansion so that the dimensional changes associated with temperature changes reflect the anisotropy of the product. Typical values for linear thermal expansion for uniaxial APC are listed in Table 1.1.

TABLE 1.1
COEFFICIENTS OF LINEAR THERMAL EXPANSION FOR UNIAXIAL APC (PER °C)

	23°C to 143°C	*143°C to 343°C*
Along fibre	$0{\cdot}5 \times 10^{-6}$	$1{\cdot}0 \times 10^{-6}$
Across fibre	30×10^{-6}	75×10^{-6}

Figure 1.2 shows representative values of density as a function of temperature. The anisotropy of thermal expansion means that it is desirable to balance the organisation of plies within a moulding if warpage is to be avoided. Even with such a lay-up the difference between the thermal contraction in layers of different direction will give rise to thermal stresses within a moulding. This phenomenon has been studied in some detail by Nairn.[15]

Our experience is that, in the initial stages of the cooling process, these stresses are readily dissipated, but that they start to lock in at about 220°C.[16] As such, the level of thermal stresses induced in APC mouldings is comparable with those in conventional high temperature curing thermosetting systems except that, as a tough matrix, PEEK does not permit stress relaxation through transverse ply cracking.

Even in a quasi-isotropic lay-up, the thermal expansion is low, indeed it is lower than that of steel: this can cause embarrassment if steel matched die moulds are used for the preparation of mouldings without adequate forethought about mould release.

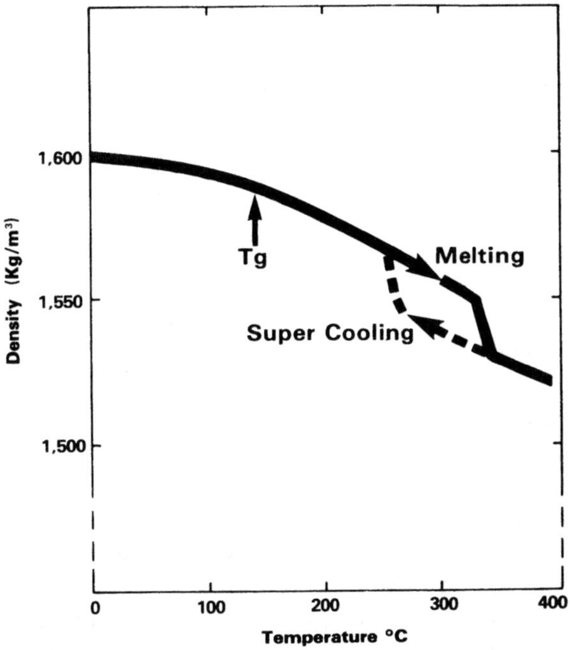

FIG. 1.2. Representative plot of density as a function of temperature for APC-2 composite material. Note that as a crystalline polymer some degree of supercooling can occur. This graph shows the expected behaviour for the recommended processing cycle based on rapid cooling.

1.4.2. Temperature, Heat and Heat Exchange

The melting point of PEEK resin is about 340°C; however, as with all semi-crystalline polymers some residual nuclei will exist to well above this temperature. The preferred melt processing temperature is 380°C with a recommended processing window of 360°C to 400°C. Below 360°C there is the possibility that the melt will not be homogeneous because of residual crystalline nuclei, while above 400°C there is the possibility of chemical change in the resin.

In respect of thermal stability APC-2 shows no significant chemical change after cooking at 400°C for 1 h, providing oxygen is excluded. The presence of oxygen, or exposure to temperatures above 400°C, can cause some changes in the resin; however, these changes have always been noted to be of a branching character, effectively increasing the toughness of the resin phase, so that mouldings made under

TABLE 1.2
HEAT CONTENT OF APC-2 RELATIVE TO 20°C

Temperature	*Heat content relative to 20°C (kJ/kg)*
20	0
50	33
100	89
150	148
200	213
250	285
300	384
350	469
400	559

these conditions do not suffer any obvious degradation of properties although some reduction in crystallinity may be apparent. Satisfactory APC mouldings have been made at 440°C.

The heat content of APC as a function of temperature is shown in Table 1.2. A representative value for specific heat in the melt is 1·8 kJ/kg°C and near to room temperature is 1·1 kJ/kg°C.

For heat exchange the concept of anisotropy must again be invoked and typical values for uniaxial material, obtained by Grove and Short,[17] are shown in Table 1.3. For most processes the thermal cycle is dominated by diffusivity normal to the fibre orientation. Axial conductivity of heat becomes significant in continuous tape processes such as filament winding or tape laying.

TABLE 1.3
COEFFICIENT OF THERMAL DIFFUSION FOR APC

Temperature (°C)	*Thermal diffusivity (m^2/s)*	
	Along fibre direction	*Transverse to fibres*
20	28×10^{-7}	$3{\cdot}4 \times 10^{-7}$
50	29×10^{-7}	$3{\cdot}5 \times 10^{-7}$
100	31×10^{-7}	$3{\cdot}6 \times 10^{-7}$
150	30×10^{-7}	$3{\cdot}6 \times 10^{-7}$
200	29×10^{-7}	$3{\cdot}4 \times 10^{-7}$
250	28×10^{-7}	$3{\cdot}3 \times 10^{-7}$
300	28×10^{-7}	$3{\cdot}2 \times 10^{-7}$
350	28×10^{-7}	$2{\cdot}7 \times 10^{-7}$
400	28×10^{-7}	$2{\cdot}4 \times 10^{-7}$

1.4.3. Crystallinity and Crystallisation

As with any crystalline thermoplastic the level of crystallinity depends on thermal history. Under very rapid cooling conditions it is possible to cool polyether etherketone below the glass transition temperature, ~140°C, before it has a chance to crystallise—under these conditions the moulding would be amorphous—conversely, if the moulding is held for a long time between 300°C and 340°C then the resin may over crystallise. Inadequate crystallisation may compromise the environmental resistance of the resin, excessive crystallisation may cause a reduction in the toughness of the composite. For APC-2 the crystallisation temperature varies with cooling rate as shown in Fig. 1.3. This leads to a qualitative interpretation in respect of crystallinity as follows:

cooling rate slower than 1°C/min: excessive crystallinity
10°C/min to 700°C/min: optimum
cooling rate faster than 1000°C/min: reduced crystallinity

This is an exceedingly wide processing window but the fact that it is a window at all can cause some anxiety. To overcome this anxiety let us consider the implications. To avoid excessive crystallisation in

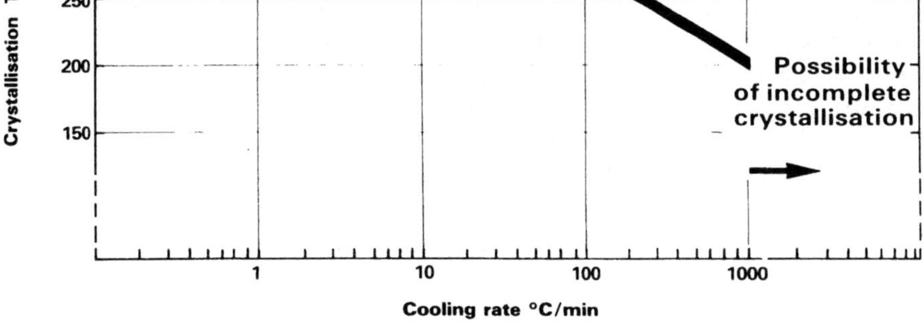

FIG. 1.3. Crystallisation temperature as a function of cooling rate for APC-2.[29]

APC-2 it is necessary to complete the cooling from 343°C to 143°C at a rate greater than 1°C/min, 200°C in less than 3 h: considering that most people interested in using thermoplastic composites will be seeking rapid processing this side of the window is unlikely to represent a constraint. In respect of rapid cooling, when we take into account factors such as surface heat transfer, it is possible to demonstrate that, even when bringing the molten moulding into contact with a heat sink at room temperature, it is impossible to pass beyond this limit for a moulding which is more than 6 mm thick. In cases where a thin moulding is subjected to intense cooling—and cooling rates of 8000°C/min have been recorded in extreme cases of stamping thin sheets into cold metal moulds—then it is always possible to anneal the moulding to an acceptable level of crystallinity by a short heat treatment at 250°C.

With respect to these recommendations on optimum crystallinity we can offer the further assurance that the disadvantages associated with falling outside this window are relative only: an excessively crystallised APC-2 moulding is still tough in comparison with mouldings from conventional thermosetting resins; and an amorphous PEEK moulding still retains superior environmental resistance in comparison with other non-crystalline high performance thermoplastic resins. The recommendation of this processing window is only a guidance to obtaining optimum properties—they are guide lines beyond which it is difficult to stray.

1.4.4. Rheology

The viscosity of the resin phase in APC-2 is high in comparison with that of thermosetting resins so that bleeding of resin from the prepreg during fabrication is only rarely observed. One exception to this is in the fabrication of thick uniaxial mouldings in a matched die mould, when some resin may be squeezed out at the ends of the moulding. Although such mouldings are important for the evaluation of properties, they rarely represent practical fabrications. The high viscosity of the thermoplastic resin plays an important role in maintaining the predetermined fibre organisation during processing so that mechanical properties measured on coupons may be translated into the performance of components. Highly viscous though the resin is, it is sufficiently mobile to permit some flow within the prepreg and also of one layer of prepreg relative to another.

Individual layers of preimpregnated tape will spread sideways under pressures of the order $10^6\,N/m^2$ (10 atm). This sideways flow is lubricated by the resin rich surface layer of the prepreg—a layer that is a few micrometres thick only. Such transverse flow is, however, limited to about 40% even when high pressure is sustained for a long time: this limitation is probably due to slight twisting of the fibres locking the deformation. The ability for some transverse flow is of particular importance when fabricating with APC prepreg tape woven into a dry fabric: the transverse flow is able to heal the gaps in such fabric, which may otherwise be exaggerated when forming double curvature mouldings.

As well as transverse flow within a ply there is also the possibility of interlaminar flow: the individual layers of prepreg can slip over each other when the material is molten. This flow is manifested as an apparent interlaminar slip whose rate is dependent on the shear stress

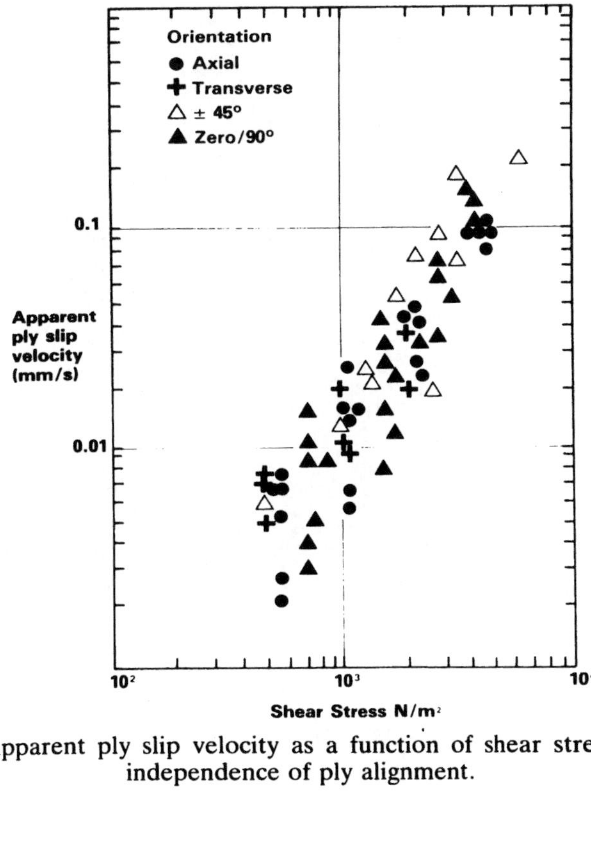

FIG. 1.4. Apparent ply slip velocity as a function of shear stress. Note the independence of ply alignment.

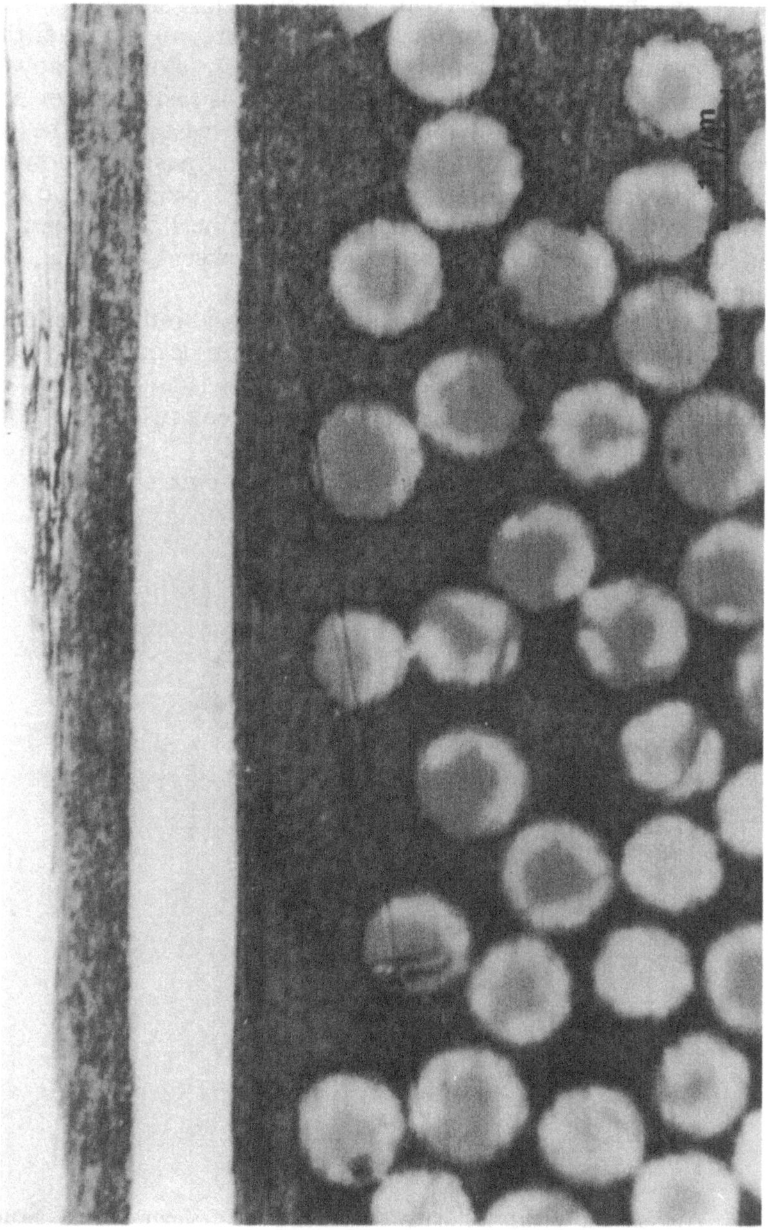

FIG. 1.5. Resin rich layer at interface between two plies of APC prepreg.

between the plies (Fig. 1.4). Although it is convenient to consider this as a slip process, our observations suggest that it is, in fact, a true viscous flow—without any discontinuity—within the resin rich layer formed where two prepregs join (Fig. 1.5).

One convenient result of the deformation process being limited to these few microns of resin rich material is that, during simple shear flows, the slip velocity does not depend on the orientation of the plies: the flow between a set of uniaxially arranged plies is the same as that between plies laid up in cross-ply organisation. An inconvenience of the constraint of the flow to such a narrow layer, formed between fibres of comparable size, is that inevitably the thickness of the layer varies and there is considerable variation in the actual slip velocity for different samples. The ability for some interply slip to take place is highly desirable when fabricating curved surfaces from consolidated sheet stock.

As well as flow in the melt valuable shaping processes can be obtained from deformations of dry fabrics woven from APC prepreg: satin weaves are especially interesting fabric forms in this respect. With the potential for combining dry fabric handling with transverse flow within the prepreg and interlaminar shear flows in the melt, to which may be added 'dodges' such as precrimping the prepreg to allow the crimp to be extracted during forming,[14] APC offers a wealth of rheological strategies in a system which, at first glance, would appear to be unduly constrained by the virtually inextensible fibres which represent 60% of the volume.

Those whose worthy preoccupation is with understanding the mechanical properties of continuous fibre reinforced plastics, and who construct models of the micromechanics of such systems, should pause for a moment to consider that the selection of resin molecular weight and fibre organisation are no less important in designing the processability of a composite than its mechanical properties. In the optimum composite both performance and processing must be optimised.

1.5. FUNDAMENTAL ENGINEERING CONSIDERATIONS

Several fundamental interactions with the shaping process must be considered in order to achieve successful fabrication of components from APC, these include: conformity to shape; air entrainment; time, temperature and pressure; and surface quality.

1.5.1. Conformability to Shape

First is the question of assuring that the fibre orientation in the final moulding is that which is required. Most mouldings will require multiple layers of prepreg (8 plies per millimetre thickness): to prevent unscheduled movement of the plies before they fuse they should be tacked together by local heating.

When preparing to mould curved surfaces designing interply slip is important in avoiding wrinkling of the inner (concave) face. Simple flanging is readily achieved in the melt state when the slip paths are short, even when sharp radii are involved. In the special case of adjacent reversed curvature, as occurs in a top hat section, Fig. 1.6,

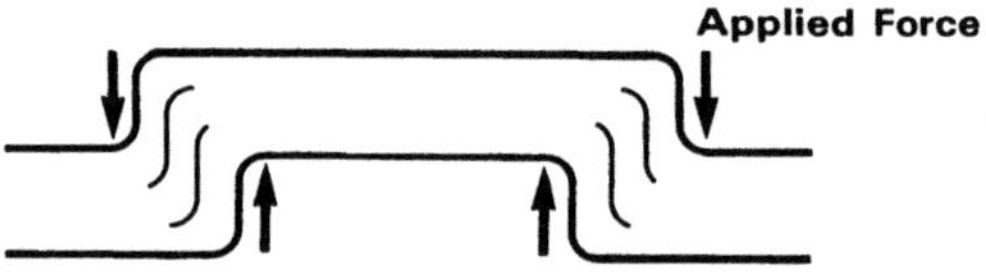

FIG. 1.6. Ply slip in 'top hat' section.

the slip surface can be contained entirely within the moulding. For the more gentle curvatures typical of aerofoil sections, Fig. 1.7, the slip paths are much longer, and are unfavourably aligned to the forming stress. Such a shallow curvature may appear simple, but even the slightest compression may induce fibre wrinkling. For such sections we recommend that the shaping phase should take place *before* melting the prepreg so that the dry, stiff, layers can readily slide over one another before or as they fuse.

In precision moulded parts note that if the mould is to experience large excursions in temperature then it will expand, or shrink, according to the thermal expansion of the metal used. Almost certainly that thermal expansion coefficient will be greater than the coefficient of thermal expansion of the composite in the plane of the moulding.

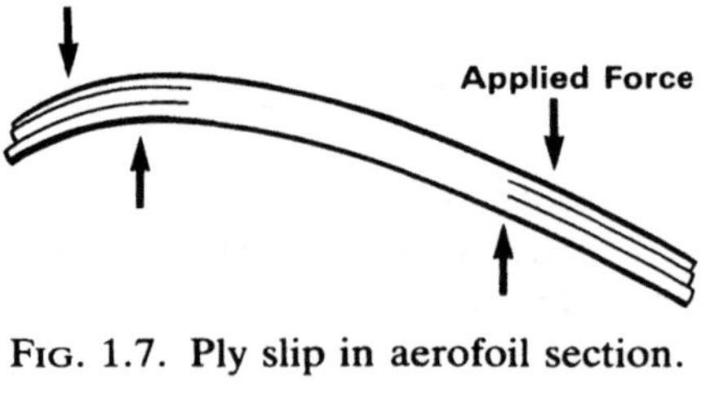

FIG. 1.7. Ply slip in aerofoil section.

1.5.2. Air Entrainment

Air entrainment can be responsible for incomplete consolidation, for voids in consolidated mouldings, for poor conformation to the mould surface, for poor surface finish, and, if processing temperature is excessively high, for oxidative attack on the resin (Section 1.4.2). Where possible vacuum conditions are to be preferred, but are not essential if proper consideration is given to venting moulds.

1.5.3. Time, Temperature and Pressure

APC prepreg will fuse in seconds at 380°C. As a preimpregnated product it does not rely on time and pressure in the melt to achieve wetting of the fibre by the resin. Nevertheless, pressure, and especially time, can be valuable allies.

Unless heat is being supplied directly to the inside of the prepreg—for example by using resistance heating based on the limited electrical conductivity of the carbon fibre—then processing will rely on conductivity of heat through the stack of prepreg plies. Although thermal conductivity within APC material is high, in the initial stages heat will be conducted through a stack of unfused plies with the additional impediment of the surface heat transfer between each layer. Quite the easiest factor to overlook in the heating stage of a stack of prepreg plies is that the amount of heat taken to melt them may be a significant proportion of the total heat content of their surroundings, which in turn must be brought back up to temperature.

We know that several fabricators working with APC have achieved satisfactory mouldings under vacuum conditions. Vacuum moulding is not a skill which we have acquired ourselves and for most fabrications we prefer to work at pressures of the order $10^6\,N/m^2$ (10 atm). Where sharply radiused curvature is concerned higher pressures may be necessary to produce the shear stresses required to achieve interply slip or transverse flow (Section 1.4.4). Pressure, like time, can be a valuable ally in the optimised forming of APC, especially with more complex mouldings.

As well as heating up a moulding process design must also consider the cooling stage. For any thermoplastic, uniform controlled cooling is preferred if an optimum product is desired: APC is no exception.

1.5.4. Control of Surface Quality

PEEK is an excellent hot melt adhesive. To avoid an APC moulding adhering to the mould a suitable mould release agent should be used.

If an especially clean surface finish is required it can be obtained by using a thin aluminium foil as the surface layer and subsequently dissolving that foil off in caustic.

1.6. LAMINATION OF FLAT SHEET

The lamination of prepreg plies or woven layers to form flat sheet is important, first, to provide coupons for mechanical property characterisation and, secondly, as a feedstock for subsequent thermoforming. There are two classes of lamination process for flat sheet: platen pressing and compliant surface moulding. With an eye to quality of the final moulding we must pay special attention to the removal of air and to the cooling of the laminate at the solidification state.

1.6.1. Platen Pressing

For conventional lamination, platen pressing requires a high degree of platen flatness and dimensional stability, coupled with pressures high enough to remove the interlaminar air. Heating and cooling under pressure are essential features, and uniformity of heat exchange is very highly desirable. These bald statements would be true of any thermoplastic compression moulding process, but they are especially relevant to processing of APC where the temperatures are high and where precision is an essential feature for such a high value product. Recommended conditions for platen pressing are reproduced in Appendix I.

ICI has developed platen pressing up to 0·75 m × 0·75 m and, as well as supplying APC in prepreg form, can also supply laminated structures. Standard forms include uniaxial sheets or quasi isotropic lay-ups $(+45, 90, -45, 0)_{NS}$, where N is a number indicating a repeated sequence and S indicates symmetry about the centre, but preferred lay-up configurations can usually be accommodated.

Improved surface quality can be obtained in platen pressing if the prepreg stack is encased in a foil package and a vacuum is drawn to remove the air from the stack before consolidation. Improved uniformity of pressure can be obtained by incorporating a compliant layer between the platen and the stack of plies.

The major advantage of platen pressing is that it is the most commonly available, and readily adaptable, of all fabrication technologies.

1.6.2. Compliant Surface Moulding

Vacuum bag moulding has been well developed for conventional thermoset composites. The basic process consists of encasing the prepreg in a flexible bag, drawing a vacuum against a moulding surface and curing the resin in a high temperature autoclave. In addition, there is the use of breather cloths to ensure that the vacuum vent paths are diffuse and effective, and the possible addition of external pressure to 'pressure assist' consolidation. We know that some people skilled in this art for epoxy composites have successfully translated this technology to operate with APC systems.

An alternative form of compliant diaphragm moulding especially suitable for APC materials has been invented. As a starting point the new technology uses the comparatively recent process for shaping metallic diaphragms known as superplastic metal forming.[18] Various metallic alloys are available, but of special interest in conjunction with APC forming is the aluminium alloy range known as Supral. These alloys have a superplastic temperature range which is compatible with APC melt forming. The process has the benefit of being originally designed to operate at the high temperatures necessary to APC, there are vacuum facilities available, the compliant diaphragm is plastically deformed under a controlled pressure and there is also scope for cooling under pressure. The first Supral compliant diaphragm moulding machine has been designed to laminate APC on the scale $2\,m \times 1\,m$.

A special type of diaphragm forming is double belt continuous lamination. This type of process has various proprietary forms but all follow an essentially similar pattern. The laminating machine is fed with layers of prepreg, these are intimately engaged by the counter rotating double belts and the plies are laminated by traversing with the belts, under sustained pressure, through a heating and subsequent cooling stage. Continuous lamination of this type is an attractive route to the economic fabrication of stock sheet for subsequent thermoforming and in particular for the long sections necessary for roll forming.

The double belt principle allows interlaminar air removal by the gradual roller type engagement of the plies, followed by sustained pressure by the compliant belts: the air is left behind at the beginning of the process. Our experience with this technology gives encouraging, but imperfect results. The temperature and cooling rate requirements present a challenge to existing machine design and further development is required to define an optimised production process.

For the longer term, compliant surface moulding strategies have three significant advantages over platen pressing: air entrainment is more easily avoided, the compliant surface obviates the need for very flat and parallel mould surfaces, and they are more readily scaled to large size.

1.7. DIRECT MOULDING FROM APC TAPE PRODUCTS

Tape laying and filament winding are two technologies which have been highly developed for thermosetting resin systems and which can, in principle, be adapted to thermoplastic systems.

1.7.1. Filament Winding

Bowen[19] has pointed to the potential for filament winding technology with thermoplastic prepreg and filament wound prototype components in APC have been demonstrated—Fig. 1.8. Filament winding with thermoplastic systems uses the same basic equipment as that which has been developed for thermosetting resins but requires an additional source of heat to fuse the tape to the substructure—Figure 1.9. For

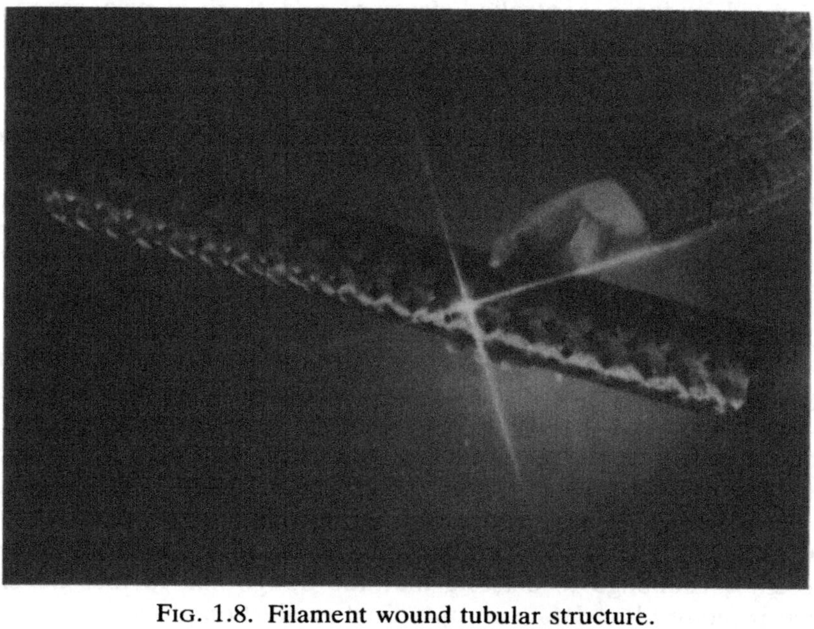

FIG. 1.8. Filament wound tubular structure.

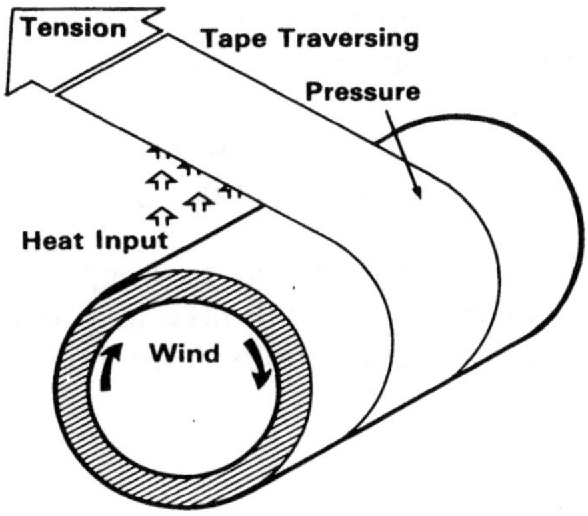

FIG. 1.9. Schematic diagram for filament winding process.

some structures it is also desirable to apply some heating to the mandrel.

It is in the use of filament winding technology that thermoplastic composites will demonstrate one of their distinct advantages. Because there is no autoclaving stage with a wet slippery system, and indeed, because the tape can be cooled immediately after it is fixed to the substructure, it is possible to consider non-geodesic winding patterns. Thus the product designer is no longer constrained from defining the optimum orientation of the reinforcement in this composite. Another advantage will be reaped from the opportunity to thermoform the component subsequently so that, for the first time, the designer will be able to envisage monolithic hollow structures with significant concave as well as convex surfaces—Fig. 1.10.

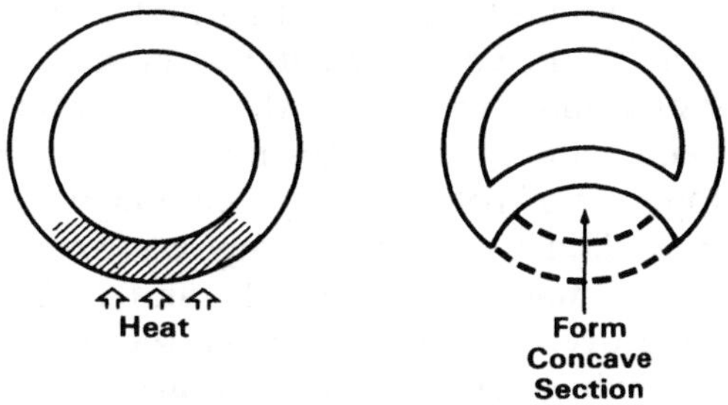

FIG. 1.10. Potential for introducing concave curvature in filament wound structures by post forming.

Filament winding has been a preferred technology for the fabrication of large structures, but, with high performance thermosetting systems the scale of the structure has always been limited by the scale of the autoclave available for the curing stage. For thermoplastic filament winding scale is limited only by the imagination of the designer.

We anticipate that, once APC single tow impregnated tape becomes freely available, filament winding will take its proper place as a major fabrication technology for continuous fibre reinforced thermoplastic composites.

1.7.2. Tape Laying

Tape laying is a big cousin of filament winding. However, because it operates with flat or low curvature surfaces, the pressure to obtain consolidation during fusion cannot be applied from tension in the tape. Instead pressure must be applied during consolidation from the tape laying head. ICI has experimented with this technology, which shows promise, but a number of engineering difficulties have to be overcome before an optimised technology can be recommended.

1.8. ROLL FORMING

While we of the plastics industry pride ourselves on the ease and economy with which parts may be moulded, we have to acknowledge that—when it comes to moulding large parts rapidly—the metal working industry has much to teach us.

One exceptionally powerful category of metal working is based on roll forming. The possibility of translating that technology to work with thermoplastic composites received its first and most adventurous champion in Lubin[20] who conceived the Grumman beam-builder for space station construction. We believe that the advent of APC prepreg will prove to be an enabling step in furthering that concept.

For our development of roll forming technology we are indebted to guidance from E. Griffin:[21] the technology was specifically developed to operate with APC materials (Figs 1.11 and 1.12) and includes an infra-red preheating oven designed to bring the preconsolidated sheet stock up to the moulding temperature, followed by a series of rolling stations to form and consolidate.

In operating with thermoplastic materials it may not be necessary to

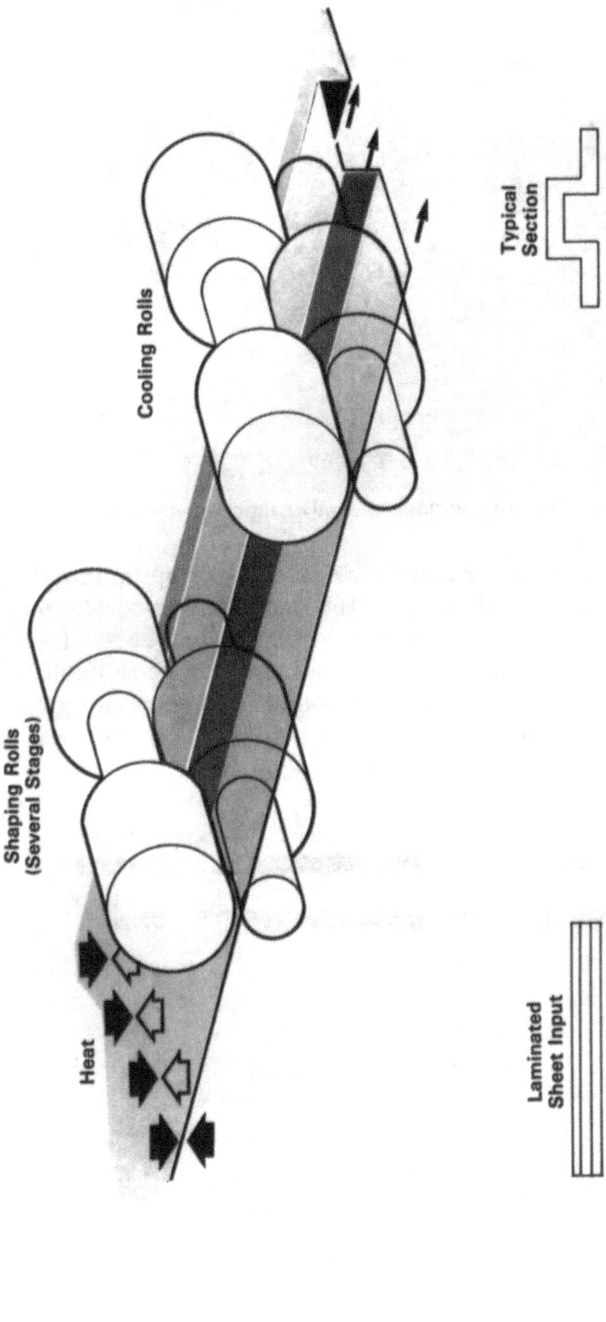

FIG. 1.11. Schematic diagram for roll forming.

FIG. 1.12. Equipment for thermoplastic roll forming.

heat the whole of the preformed blank to achieve optimised shaping. Thus to form a 'top hat' section only the material which will end up in the corners and side walls needs to be heated—Fig. 1.13. For convenience of material handling it may well be sensible to retain rigidity in the preform, and this is obtained by screening part of the blank during the preheating stage.

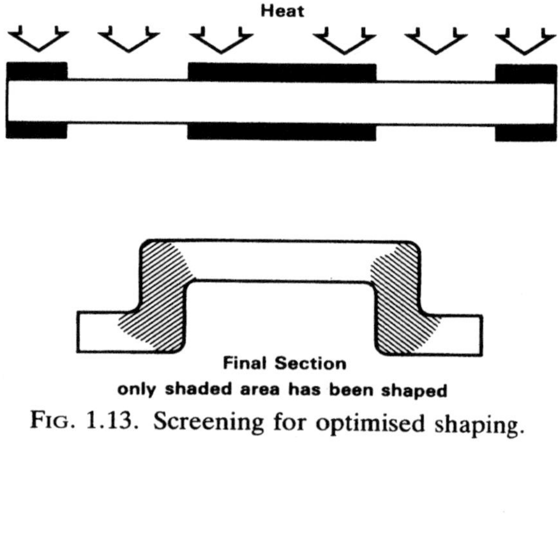

FIG. 1.13. Screening for optimised shaping.

FIG. 1.14. Various roll formed sections.

FIG. 1.15. Roll formed corrugated panel.

FIG. 1.16. Interply slip in channel flange.

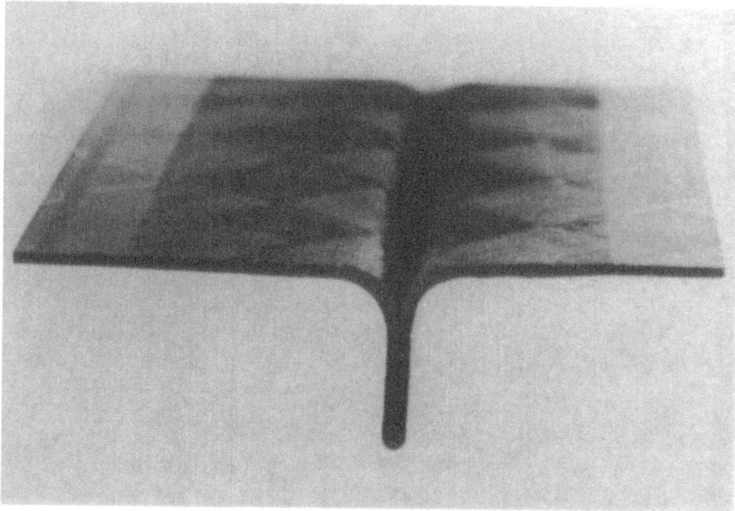

FIG. 1.17. Interply slip in 'tee' section.

Because roll forming is a precision technology, it requires a precision feedstock—especially in respect of section thickness. The technology developed in-house by ICI has been used to fabricate a variety of sections—Fig. 1.14—at speeds up to 10 m/min. These examples include curved sections and sections profiled in the machine direction—Fig. 1.15.

To demonstrate shape in roll forming is only a first requirement. For a useful structural component we need to demonstrate that consolidation is retained in the final product and that there has been no ply buckling on the internal concave sections. Figure 1.16 shows a channel flange section roll formed in APC showing interlaminar slip which accommodates the deformation without ply buckling. Here the slip plane has been sheared to the edge of the moulding. In Fig. 1.17 we see the whole of the interply slip accommodated in a double reverse bending to form a 'tee' section. Full details of the ICI roll forming process are available.[22]

1.9. COMPRESSION FORMING

The most prolific of modern metal technologies is sheet working. Of the various options for sheet working hydroforming[22] is an attractive option for APC fabrication.

As with roll forming there is only the need to supply an appropriate heating source to preheat the APC blanks and conventional cold metal working technology can be immediately adopted. Our demonstrations of hydroforming, Fig. 1.18, were carried out under guidance from T. G. W. Stephens.[23] As with metal working one of the chief skills of hydroforming APC lies in the cutting of the blanks and selection of appropriate tooling: male tool (when the blank is formed around the outer surface of the die or punch); cavity tool (where the blank is formed against the inner surface of the mould); or double-acting tool (which requires a press that closes in two stages, first to hold down the perimeter of the blank and then second to engage a punch to form it). Figure 1.19 demonstrates a range of hydroformed sections, most of which are relatively simple folded structures, but which include sections with two dimensional curvature, like the Rolls Royce blister fairing.[24]

Our experience with APC has been limited to presses up to half a metre diameter, but we understand that fluid form presses 1 m in diameter are available.

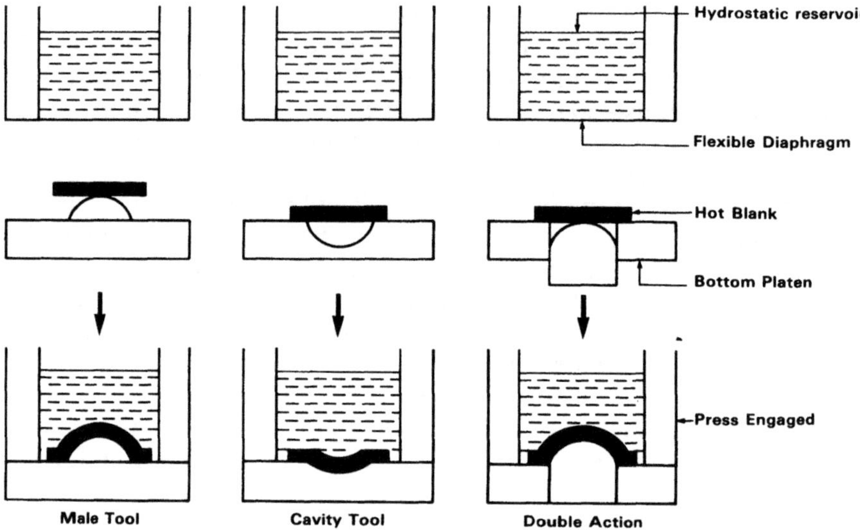

FIG. 1.18. Schematic diagram of hydroforming.

Hydroforming is only one of a range of rubber forming processes: rubber pressing is another example, based on a solid rubber block rather than a pressurised diaphragm. At first, the fact that rubber tooling can be advocated at all appears an anomaly when we consider that, to be successfully formed, the blank must be at 380°C. However, the temperature of the blank is only one factor in the equation, the thermal capacity of the tooling being a second important issue. Our experience is that rubber tooling with the comparatively low continuous service temperature of 150°C can give satisfactory performance[25]—indeed we have used wooden tooling.

The high thermal capacity of the tooling in forging processes can lead to exceedingly rapid heat transfer: cooling rates in the critical region 340°C to 140°C in excess of 5000°C/min have been recorded. Such cooling rates can lead to reduced crystallinity in the moulding (Section 1.4.3). When such rapid cooling has been used it is appropriate to use a subsequent annealing process to develop the optimum crystallinity, but since this requires a second operation, it should be avoided if at all possible. The best way to avoid excessive cooling is to use warm tooling, and the best possible temperature for the press is about 180°C.

The option of utilising well established metal working technologies

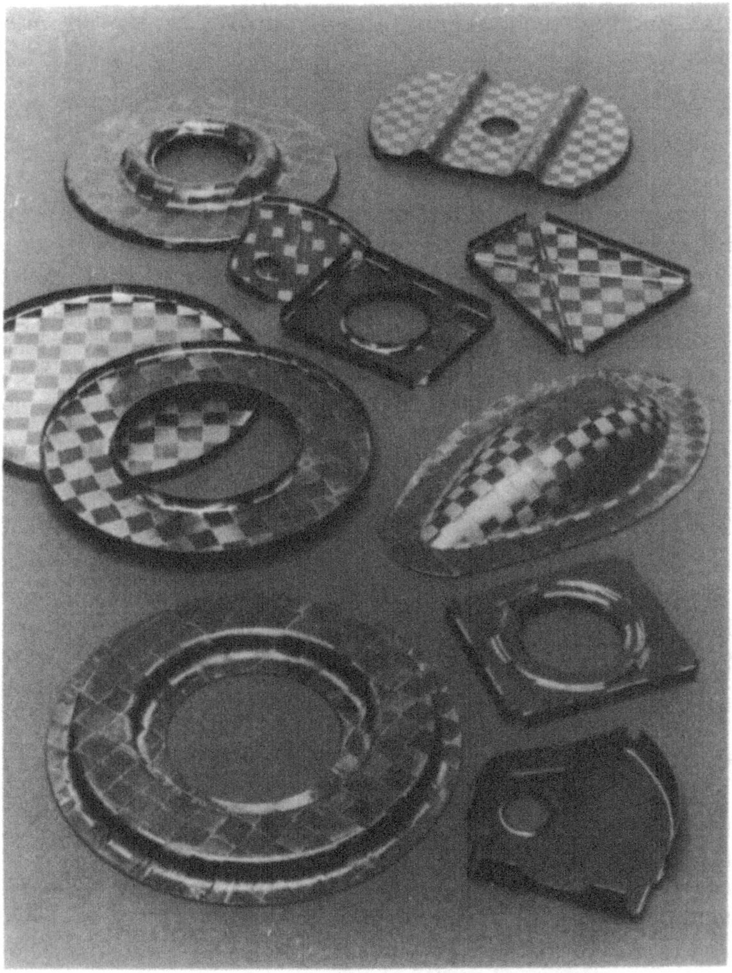

FIG. 1.19. Various hydroformed shapes.

such as hydroforming to process advanced composite materials such as APC should prove to be a highly attractive option for established metal working companies. While some people may mourn the loss of the traditional composite fabrication technologies of wet lay-up followed by autoclave curing when they move to thermoplastic composite systems such as APC, the opportunity to view the latter product as black aluminium to which they can apply the whole range of metal bashing technologies should prove to be an adequate compensation.

1.10. DIAPHRAGM FORMING

One of the most promising technologies which we have developed for the fabrication of large area double curvature formings is based on an extension of superplastic forming in conjunction with Superform Metals Ltd.[18]

To fabricate complex curvative sections with this technology, the APC tape is first assembled in a desired lay-up or woven form and encased, between thin Supral aluminium sheets. A vacuum is then

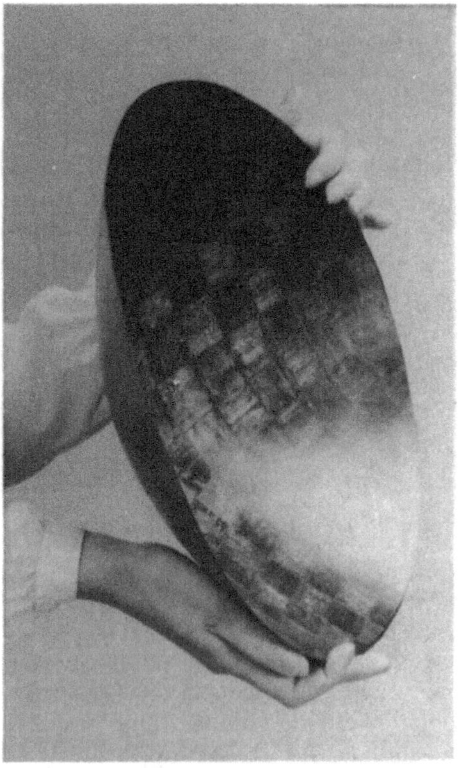

FIG. 1.20. Parabolic double curved section.

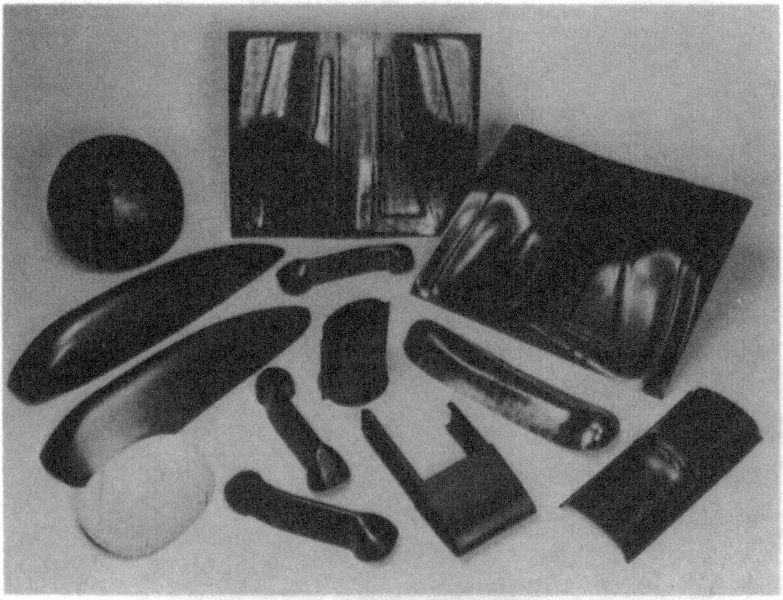

FIG. 1.21. Components fabricated in conjunction with Superform Metals Ltd.

FIG. 1.22. Ducting inlet fabricated in APC.

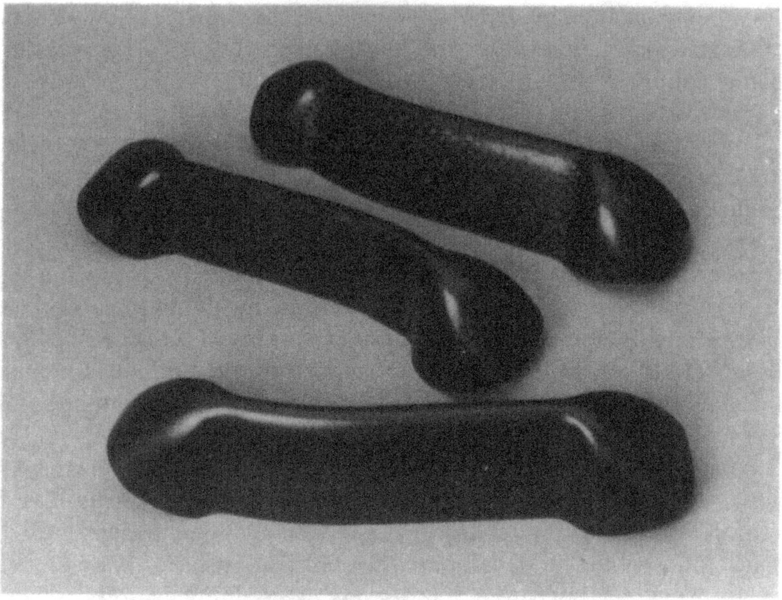

FIG. 1.23. 'Dogbone' component fabricated in APC.

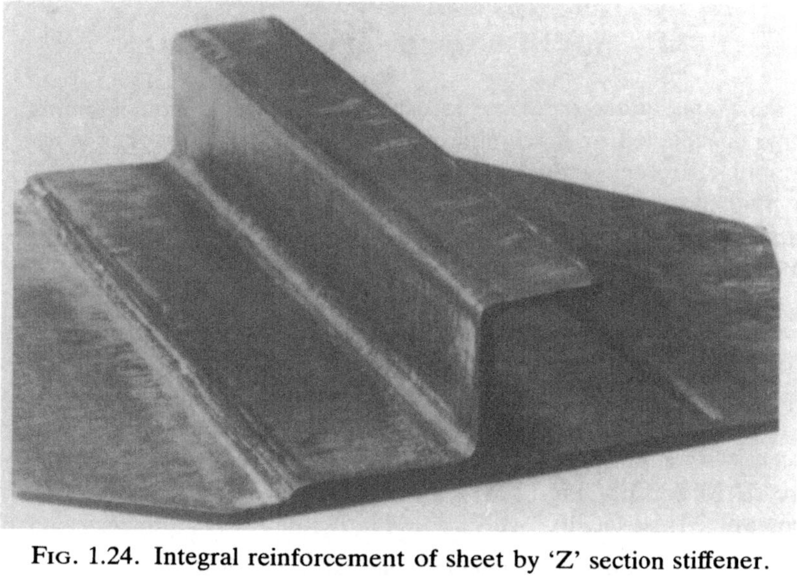

FIG. 1.24. Integral reinforcement of sheet by 'Z' section stiffener.

drawn between the sheets. The ambient temperature is now raised to 380°C, at which temperature the Supral sheets become superplastic so that the sandwich structure can be shaped by differential pressure into an appropriate mould where it is cooled.

A more complete description of this process, which is the subject of patent applications, will be presented elsewhere.[26]

Figure 1.20 demonstrates a superplastic diaphragm formed paraboloid section having a well defined double curvature. The variety of sections illustrated in Figs 1.21 and 1.22, formed by this technology, will be recognised by many as difficult even in aluminium, while the very challenging demonstrator components of Fig. 1.23 suggest that the full versatility of the technology has yet to be defined. Integrally reinforced structures can also be fabricated by this technique—Fig. 1.24.

Superplastic forming is a new technology in metal working. We believe that it will prove to be an especially powerful technology for thermoplastic composite materials. Like many other branches of the plastics industry in the development of fabrication technology for APC systems we have borrowed extensively from the best process technologies available for other materials. This is not necessarily a one way traffic: the development of superplastic forming for APC may lead to improved superplastic forming for metals.

1.11. MACHINING OF APC COMPOSITES

In the manufacture of many aerospace components from aluminium, shape is achieved by machining from a block. Machining has not been a popular strategy with thermosetting composite materials because of the high waste, and because the brittleness of those systems makes machining difficult. We have found that the tough resin system allows APC to be machined with conventional metal working tooling, although, because the material is thermoplastic, cooling water should be applied as the cut is taken. Figure 1.25 shows typical sections machined from APC systems. While such machining will only rarely be applicable as a major shaping process, it can well play a valuable role in trimming mouldings prepared by other techniques. The ease with which APC mouldings can be machined allows holes to be drilled and threads to be cut—Fig. 1.26—without causing undue damage to the structure. This facility will undoubtedly be of significance in the development of jointing strategies.

FIG. 1.25. Components machined from APC.

FIG. 1.26. Machined screw threads and bolt holes. Note also the machined knurling on the screw head.

1.12. SCRAP RECOVERY BY INJECTION MOULDING

Victrex PEEK resin is well established as an injection moulding material and the range of commercial products available includes grades reinforced with short glass and carbon fibres.[27] These reinforced injection moulding grade resins are high value materials in their own right, offering excellent environmental resistance combined with high service temperature and stiffness for this class of mouldable resin. In respect of the ability to mould complex shapes with high precision and rapidity injection moulding has no rival amongst processing technologies.

Not surprisingly we find that scrap material from APC mouldings can be ground up, diluted with additional resin, and injection moulded with equal facility to its cousin designed specifically for that purpose. Surprisingly, however, it is possible to obtain better mechanical properties when moulding from such APC scrap than with the compound designed for that purpose. Noting this phenomenon Folkes[28] suggests, and we concur, that this is due to the compound based on APC material containing longer fibres than those of the standard compound.

The retention of longer fibres in injection mouldings based on APC feedstock is a reflection of the excellent wetting of the fibres by the viscous resins at the prepreg stage. The viscosity of the resin restricts the individual fibres from abrading each other.

Of course to use virgin APC as a feedstock for injection moulding would be an expensive option. However, technologies like sheet forging which require careful cutting of the blanks will inevitably produce some trimmings and such material, which would otherwise be scrap, can be reclaimed as a high value feedstock for injection moulding.

As well as simple injection moulding operations, the unique compatibility of the material in continuous fibre form and as injection moulding feedstock encourages the option of placing a continuous fibre reinforced element in the mould, and then injection moulding the short fibre reinforced resin around it, producing thereby the typically complex shapes of injection mouldings but selectively reinforced with continuous fibre.

The opportunity to utilise scrap material in this way, broadens the range of structures which can be moulded, and inevitably reflects on the total economics of processing with these high value composite materials.

1.13. ENVOY: THERMOFORMING PRACTICE AND THE ACHIEVEMENT OF QUALITY

Aromatic Polymer Composites have been developed to satisfy some of the most demanding structural applications yet devised. While the basic mechanical properties of this composite[3–8] family excite appreciation, translating these properties into practice demands quality in the moulded article.

Any fabricator achieves quality by a continual awareness of the processing principles involved, allied to much practical experience. It is appropriate, then, to conclude this chapter by reflecting on some of the basic rules for the successful thermoforming of APC. In doing so we shall note how those principles overlap with those already established for metal working and thermoset technologies.

Forming flat, inextensible, material into a more complicated shape requires an ability to compress or shrink the blank without it wrinkling or buckling, combined with the ability to stretch it without tearing. In general it is the prevention of buckling which requires most attention: it is essential to minimise the in-plane compression. Fortunately, certain strategies, which are well established in the compression moulding of metals, can be applied to APC forming. For example, in some technologies metal workers counter wrinkling by using mould and blank designs which restrain the forming at critical points by introducing local tension. Similarly they counter thinning or tearing by reducing clamping to allow the blank to draw or slip.

A second strategy available to APC fabrication is to replace compression by interlaminar slip. A fully molten APC laminate is a potentially slippery sandwich but interlaminar slip can only be induced by in-plane shear forces which requires appropriate design of the process tooling. Specific means of inducing shear include: crowned or plug shaped forming pads, bubble-type diaphragm contact, moulding between supportive diaphragms, and wiper rolls (especially appropriate for forming the deep curvatures of such sections as leading edges). Of course, excessive ply slip or fibre movement may not be acceptable to the product designer. It is essential that designer and fabricator jointly define and resolve any incompatibility.

Close collaboration between designer and fabricator has been an essential feature of thermoset moulding practice involving the judicious use of snipping, notching and cajoling of wet prepregs to achieve an acceptable but practical fibre arrangement. Equivalent operations may be applied with APC systems, noting that while the dry 'boardy' prepreg restricts somewhat the freedom of wet lay-up, the ability to construct dry woven fabrics may offer the innovative fabricator a whole new field in which to establish his special skills: one need only reflect on the range of designs of straw hats to stimulate the imagination.

A quality APC component should be fully consolidated, free from voids and have the surface character required by the designer. Full consolidation demands the removal of air by efficient venting combined with appropriately sustained pressure; secondly, proper precautions (noted above) to avoid wrinkling during forming must be introduced, otherwise such wrinkling can induce delaminations on the concave surface; thirdly, when using a preconsolidated blank any prestressed fibre arrangement in the blank should be minimised, since relaxation of the stresses during remelting may induce delamination; finally, premature freezing of the surface layers while shaping is still taking place must be avoided if the surface layers are not to be blistered—in such cases warm, well vented tooling is indicated. Warm, well vented tooling is equally important in ensuring freedom from surface voids or depressions.

Enhanced surface quality also requires adequate tool quality and matching, and reducing friction or adhesion between APC and the forming surface. Parallelism is a first requirement, so that pressure application is uniform. Wear, on surfaces which should be matched, should be avoided lest resin, or even composite, be squeezed into narrow gaps where it may freeze prematurely constraining the maintenance of pressure on the moulding proper during the full moulding cycle. Obviously the quality of the mould surface itself is of major importance, and a smooth release from that surface is essential. Here the judicious use of a mould release agent is desirable.

In the pursuit of quality in mouldings there is no substitute for practice. We can only offer these notes as guidance for the emerging technologies of fabrication with continuous fibre reinforced thermoplastics, hoping that they will allow those who will develop this field to exploit the opportunities which exist in linking the traditional technologies of wet lay-up with thermosets with plastic forming of metals.

ACKNOWLEDGEMENTS

We are indebted to our colleagues of the APC team and the Polymer Science Community at Wilton whose research this text reflects. We are also particularly grateful to those members of the thermoset and metal processing communities, especially A. J. Barnes of Superform Metals Ltd, E. Griffin of Kingspan Rolled Sections Ltd and T. G. W. Stephens of Cheltenham Tool Company, who have patiently shared their skills with us thereby accelerating our own learning processes.

REFERENCES

1. Belbin, G. R., Brewster, I., Cogswell, F. N., Hezzell, D. J. and Swerdlow, M. S., Carbon film reinforced polyether etherketone, *SAMPE Conference*, Stresa, 1982.
2. McMahon, P. E., Thermoplastic carbon fiber composites, in *Developments in Reinforced Plastics—4,* Applied Science Publishers, London, 1984.
3. Cogswell, F. N., Continuous fibre reinforced thermoplastics, in *Mechanical Properties of Reinforced Thermoplastics* (D. W. Clegg and A. A. Collyer, eds), Elsevier Applied Science Publishers, London, 1986.
4. Kempe, G. and Krauss, H., *Processing and Mechanical Properties of Fiber Reinforced Polyether Etherketone (PEEK)*, DFVLR, Stuttgart, 1984.
5. Bishop, S. M., The notch sensitivity and impact peerformance of carbon fibre reinforced PEEK, *RAE Farnborough Proc. Int. Symp. on Carbon Fibre Composites,* WIM, Erding, 1984.
6. Hartness, J. T., An evaluation of polyether etherketone matrix composites fabricated from unidirectional prepreg tape, Univ. of Dayton, *29th National SAMPE Symposium,* 3–5 April, 1984.
7. Cogswell, F. N. and Leach, D. C., Continuous fibre reinforced thermoplastics: a change in the rules for composite technology, *Plastics and Rubber Processing and Applications,* **4** (3), 271 (1984).
8. Leach, D. C. and Moore, D. R., Toughness of aromatic polymer composites, *Composites Conference,* Delaware, 1984.
9. Martin, C. C., O'Connor, E. and Lou, Y., Polyphenylene sulphide high performance composites, *29th National SAMPE Symposium,* p. 753, 1984.
10. Poly(amide-imide) graphite composites, Development Information Sheet, Amoco Chemicals Corporation, 1984.
11. 'Spiflex' Composites, SPIE Battignolles, 1984.
12. Gibbs, H. H., K-polymer: a new experimental thermoplastic matrix resin for advanced structural aerospace composites, *29th National SAMPE Symposium,* p. 1073, 1984.
13. 'Fibre Imprégnée Thermoplastique (FIT)' by ATO CHEM, Composites et Noveaux Materianx, 5 May, 1984.
14. Fibre Reinforced Product of Improved Formability, UK Patent Application No. 8227333.
15. Nairn, J. A. and Zoller, P., Matrix solidification and the resulting residual thermal stresses in composites, *J. Mat. Sci.,* **20,** 355–67 (1985).

16. GRAY, M. J., ICI, private communication, 1983.
17. GROVE, S. and SHORT, D., Plymouth Polytechnic, private communication, 1984.
18. BARNES, A. J., Superform Metals Ltd, Worcester, UK (a member of the Alcan Group), private communication, 1983.
19. BOWEN, D. H., Filament winding in the 1980's, *Fibre Reinforced Composites '84*, Liverpool, 1984.
20. POVEROMO, L. M., NUENCH, W. K., MARX, W. and LUBIN, G., Composite beam builder, *SAMPE Journal*, p. 7, January 1981.
21. GRIFFIN, E., Kingspan Rolled Sections, Halesowen, Birmingham, private communication, 1982.
22. ICI plc APC Data Sheets, Welwyn Garden City, Hertfordshire, 1984.
23. STEPHENS, T. G. W., Cheltenham Tool Company, Cheltenham, private communication, 1982.
24. OWENS, G. A. and LIND, D. J., Application of carbon film reinforced thermoplastic (PEEK) composite to aero engine component manufacture, *Proceedings of Fibre Reinforced Composites '84*, Liverpool, Paper 12, 1984.
25. BREWSTER, I. and CATTANACH, J. B., Engineering with long fibre thermoplastic composites, *SAMPE Conference*, London, 1983.
26. BARNES, A. J. and CATTANACH, J. B., Advances in thermoplastic composite fabrication technology, *ME! Materials Engineering Exhibition and Conference*, Nov., 1985.
27. 'Victrex' PEEK, ICI plc, Welwyn Garden City.
28. FOLKES, M. J. and KELLS, D., Fundamental considerations in the injection processing of fibre reinforced thermoplastics, *Proceedings of Fibre Reinforced Composites '84*, Liverpool, Paper 15, p. 6, 1984.
29. BLUNDELL, D. J. and OSBORN, B. N., Crystalline morphology of the matrix of PEEK–carbon fibre aromatic polymer composites—Part II. Crystalline behaviour, *SAMPE Quarterly*, Oct., 1984.

APPENDIX I: PLATEN PRESSING OF APC

1. The prepregs to be pressed should be laid up in the required orientation sequence. A hot (240 V, 175 W) soldering iron can be used for tacking the plies together as a unit. For accuracy and handling convenience the unit should be limited to eight plies. Thus several units together may make up the final stack for lamination.
2. The ply stack should be placed between high temper aluminium foils, pretreated with a surface release agent, e.g. Frekote FRP.
3. To stop material flow at the laminate edges either a matched mould or picture frame may be used. The mould faces should be able to continue to move closer together as the thickness of

the stack reduces under pressure, a feature which will avoid any misjudgement of the optimum frame thickness. Where a frame is used the foil-faced ply stack should be placed between two mirror-finish stainless steel glazing plates (minimum 0·064 in. thick).

4. The pressing thermal cycle has three distinct features, i.e. heating, consolidating and cooling. Heating of the stack requires platens with measured surface temperatures between 370 and 390°C. At a nominal contact pressure of 10 psi the required heating time is 1 min per ply up to a limit of 30 min where a picture frame is used. Consolidation is achieved by applying a pressure (usually between 200 and 300 psi) to the heated stack for 5 min. Cooling should then proceed at the consolidating pressure until the laminate matrix is cold enough to allow release and still maintain consolidation. For lamination, ICI has cooled from a nominal 380°C down to a release temperature of 200°C at a cooling rate of 40°C/min. This has been achieved by transferring the consolidated laminate from a hot (380°C) to a cold (180°C) press for the cooling. The cooling time is about 5 min.

In developing the above platen pressing technique for laminates up to 30 in. × 30 in., ICI has achieved quality via special attention to platen flatness, optimum picture frame sizing and means of cooling under reasonably sustained pressure.

Chapter 2

ADVANCES IN THE CHEMISTRY AND APPLICATIONS OF BISMALEIMIDES

D. LANDMAN

Consultant, Monterey, California, USA

SUMMARY

This chapter considers the demand for improved high performance matrix resins for composite materials, especially for aerospace. The chemistry of polyimides and of bismaleimides is outlined, and the processing of bismaleimides is described in some detail. Formulated bismaleimide composites for specific applications are discussed. The resins have considerable versatility and are used for such applications as sealants and adhesives besides fulfilling the role of the matrix in composites. Some formulations are designed for optimum moisture resistance and others are designed to withstand microcracking. The potential of bismaleimides in the electrical and electronic industries is considered. The relatively good heat resistance of bismaleimides makes them an attractive proposition in these fields. Formulations are complex and a wide range of properties and processing characteristics can be achieved.

2.1. PRESENT STATE OF THERMOSET RESIN USAGE FOR MODERATE TEMPERATURE APPLICATIONS

Epoxy resins are currently by far the largest segment of addition cured thermosetting polymers which are used to make structural composites in the aerospace industry. The reason for their wide usage is that epoxies are easy to handle and process. Epoxies used in composites can be cured at low temperatures and pressures, e.g. 250–350°F (121–177°C) at 70–100 psi (5–7 bar), which means currently available fixtures such as autoclaves rated to these conditions are usable. In most aerospace and other engineering applications, temperatures and pressures exceeding these values make it difficult to process composites.[1]

The problem with epoxies is that their upper temperature range for structural performance is approximately 350°F (177°C). Up to 300°F (149°C) performance is possible for epoxies exposed to moist environments, which limits the applications where they can be used to those where relatively non-thermally critical structural parts are required.[2–5]

Recently, interest has developed in making parts to be used closer into the thermally critical areas of aircraft such as engines. In addition, considerable interest has developed in the use of higher performance composites in military aircraft, where skin temperatures can be somewhat higher than epoxy resin capabilities. Figure 2.1 shows approximate use temperatures as a function of service life for several aerospace adhesive applications.[6] Similar conclusions are applicable to

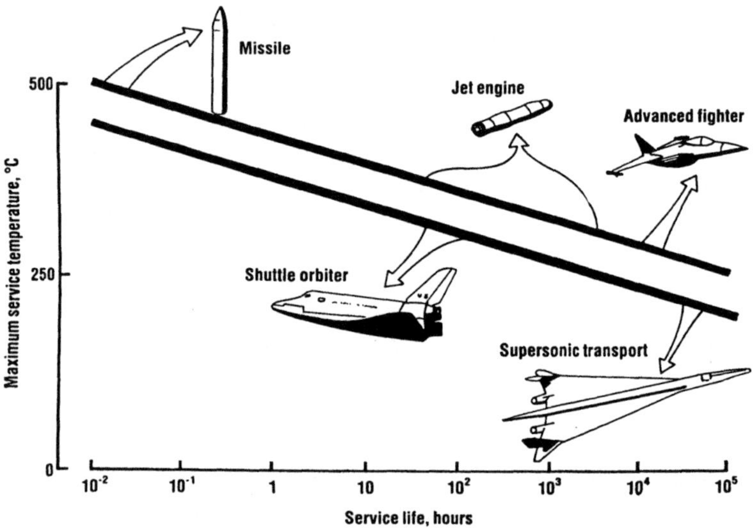

FIG. 2.1. Approximate use temperature as a function of service life for several aerospace adhesive applications. (Courtesy of *Materials Engineering*, Penton/IPC, p. 37, August, 1983.)

composite matrix resins from which it is clear that epoxy composites would be somewhat limited beyond approximately their use in advanced fighter construction.

Higher temperature performance resins are needed for composites to be used where epoxies cannot be used, but unfortunately as high temperature properties are increased, ease of processability of the resins is reduced. Many technologists involved in this problem believe that polyimides* are probably the resins of choice to meet the economic and thermal requirements[1] for higher performance materials.

2.2. POLYIMIDE CHEMISTRY AND PROCESSABILITY

Two types of polyimides are of importance viz. those produced by (1) condensation cure and (2) addition cure.

2.2.1. Condensation Polyimides

Figure 2.2 illustrates the typical chemistry involved in making condensation cured polyimides.

In the first step (Fig. 2.2, eqn (2.1)) a multifunctional aromatic amine and an acid/ester (produced from esterification of a dianhydride) are reacted to form a polyamic acid intermediate which is soluble in convenient solvents such as dimethylformamide, dimethylacetamide or *N*-methylpyrrolidone. At this point, the resin is handleable and can be used to impregnate fibres. In the following step (Fig. 2.2, eqn (2.2)) the crosslinking cure occurs to give a polyimide with release of water. Since both water and the solvent need to be driven off during the cure process, numerous voids are present in adhesive bonds and composites made with condensation polyimides. In addition, the temperature at which polyimide ring formation occurs (480–570°F (249–299°C)), as well as the pressures required to avoid excessive void formation (about 250 psi (17 bar))[7] prohibit use of conventional autoclaves.

* For the purpose of this chapter, thermoplastic polyimides will not be considered since their processing is different to that for thermosetting materials. 'Polyimide' will mean 'thermosetting polyimide' in this discussion.

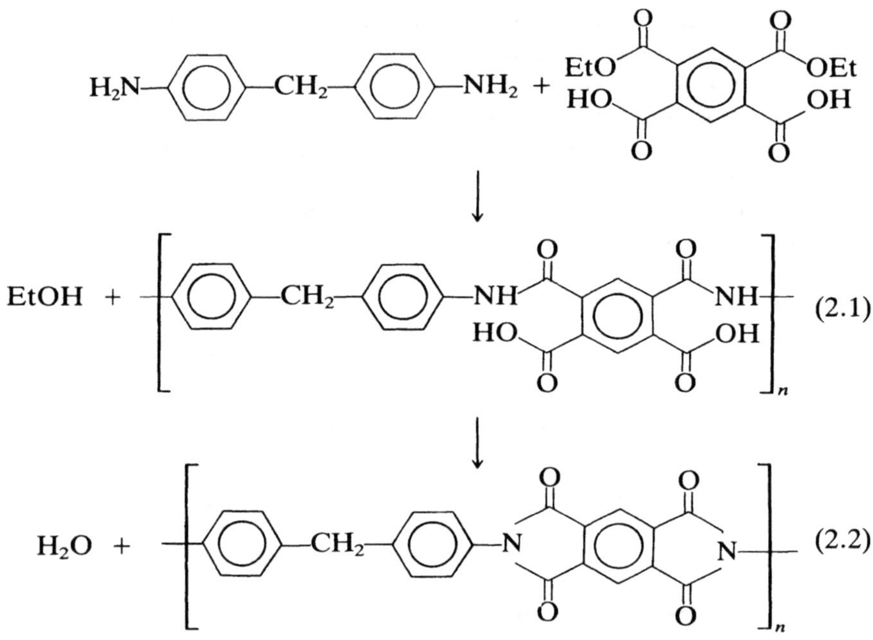

FIG. 2.2. Typical chemistry of condensation polyimides.

2.2.2. Addition Cured Polyimides

A way for condensation polyimides to be made processable has been shown above through the use of soluble polyamic acids. An alternative approach to processable polyimides is to make up a polyimide oligomer with reactive end groups which crosslink on curing. The oligomer is made by the condensation route, but the crosslinking occurs by addition cure from which no volatiles are released to cause voids.

Figure 2.3 illustrates an addition cured polyimide of the PMR,[8,9] LARC[10] or P13N[11,12] type. Here, a polyamic acid oligomer (Fig. 2.3, eqn (3.1)) is made, in which the end-cap groups (norbornene) can be later used for crosslinking by addition cure (Fig. 2.3, eqn (3.3)). Recently it has been suggested that the intermediate oligomer is a polyamide ester rather than a polyamic acid.[13] Typically, this oligomeric intermediate (Fig. 2.3, eqn (3.1)) is dissolved in ethanol or some other suitable solvent and this solution is used to coat fibres or fabric. Upon heating to 212–400°F (100–204°C), a condensation reaction

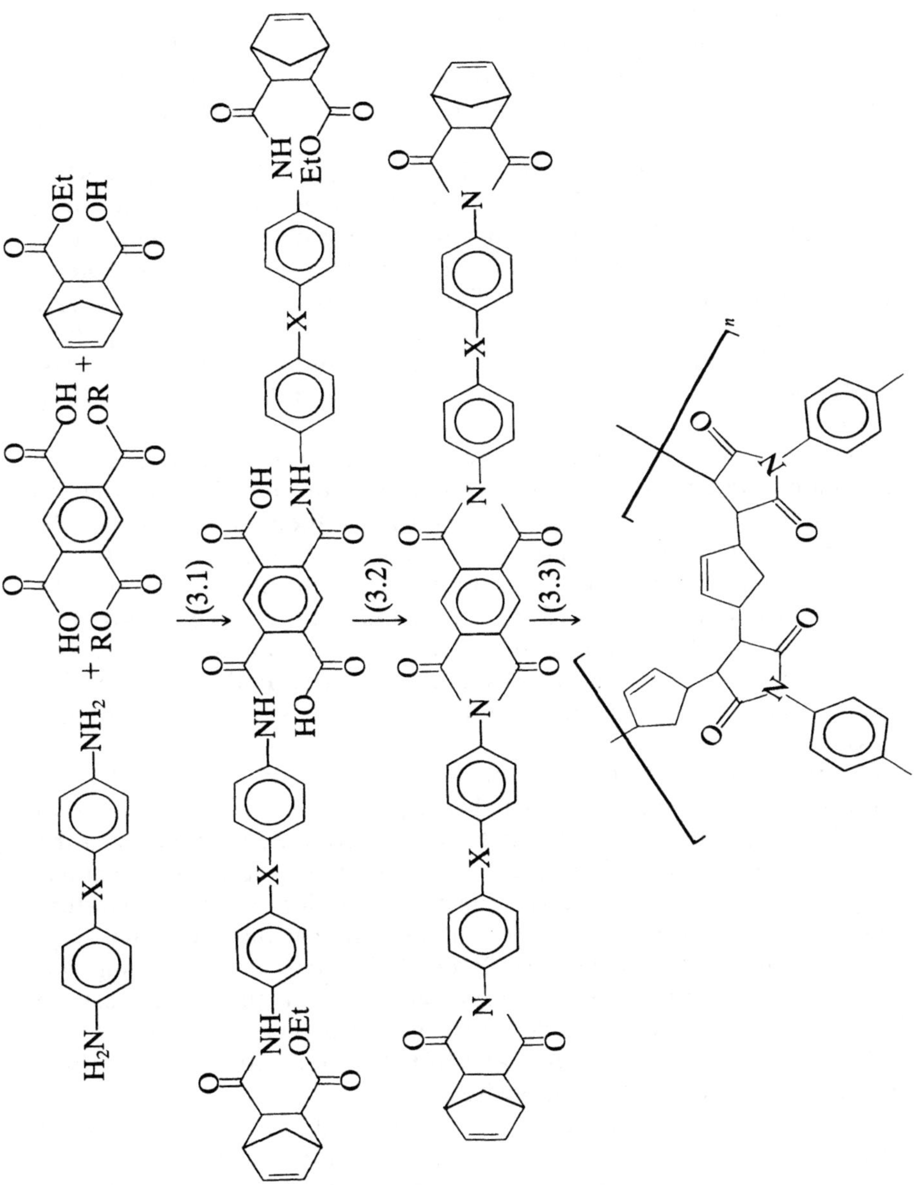

FIG. 2.3. Typical chemistry of addition cured polyimides.

releasing both water and the solvent occurs to give an oligomer of the type shown in Fig. 2.3, eqn (3.2).

The tape or fabric prepreg made by this method is usually quite dry and has poor draping and handling characteristics. Some residual solvent may be left in the resin to improve on these handling problems.

Crosslinking of these resins (see Fig. 2.3, eqn (3.3)) is accomplished by heating them to 525–600°F (274–315°C) with the application of at least 100–200 psi (7–14 bar). This high pressure requirement exists because it is well-known that norbornene rearranges via retro Diels–Alder reaction to give cyclopentadiene which would volatilise and give considerable voids.[14–16] Accordingly, an alternative end group (viz. acetylenic[17]) has also been used for making addition curing polyimides, which avoids the problem of volatiles release by end-cap decomposition.

Figure 2.4 shows the basic strategy used for making acetylene terminated polyimides. Again, a condensation reaction is used to make a polyamic acid precursor which is then converted to the oligomeric polyimide. The latter polyimide, known commercially as a Thermid 600 resin, is dissolved in solvents such as *N*-methylpyrrolidone and used to impregnate fibres. The solvent is flashed off and the prepreg is heated to approximately 375–390°F (191–199°C), at which point resin flow occurs. Full crosslinking is accomplished at approximately 660°F (349°C) with 200 psi (14 bar) pressure.

Recently, an alternative to the above acetylene terminated polyimide has been described, in which the polyimide component consists of isoimide groups.[18] It is claimed that the isoimide rearranges to the normal imide as shown in Fig. 2.5. The isoimides, described commercially as Thermid IP 600 Series resins, allow for lower temperature processing and are apparently soluble in various solvents, e.g. acetone/toluene, cellosolve glycolethers, dioxane, etc. By using solvents fibre or fabric can be impregnated and then the solvent can be flash evaporated. The isoimides melt at approximately 330°F (165°C) and apparently are fully cured at 450°F (232°C) and 150 psi (10 bar) pressure to give excellent composites.

Obviously, for all of the above addition cured polyimides there exist processing problems such as: use of solvents which need to be evaporated, use of excessive pressure or use of excessive temperature. These problems limit or prohibit the use of these addition cured

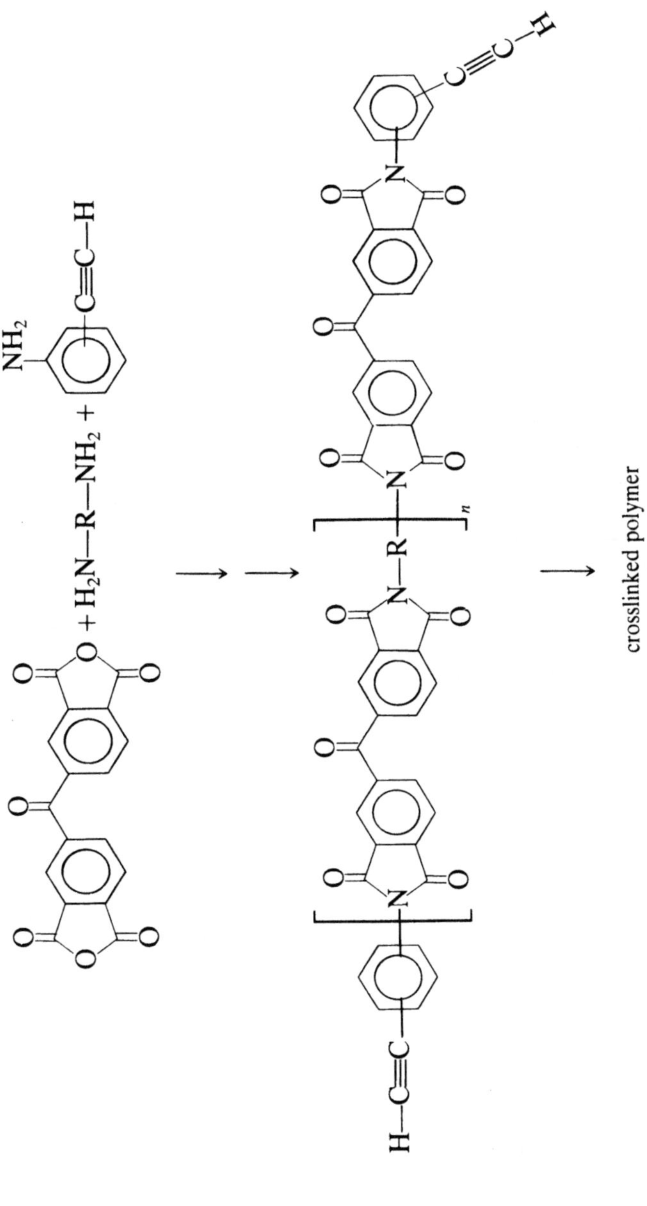

FIG. 2.4. Typical chemistry of acetylene terminated polyimide.

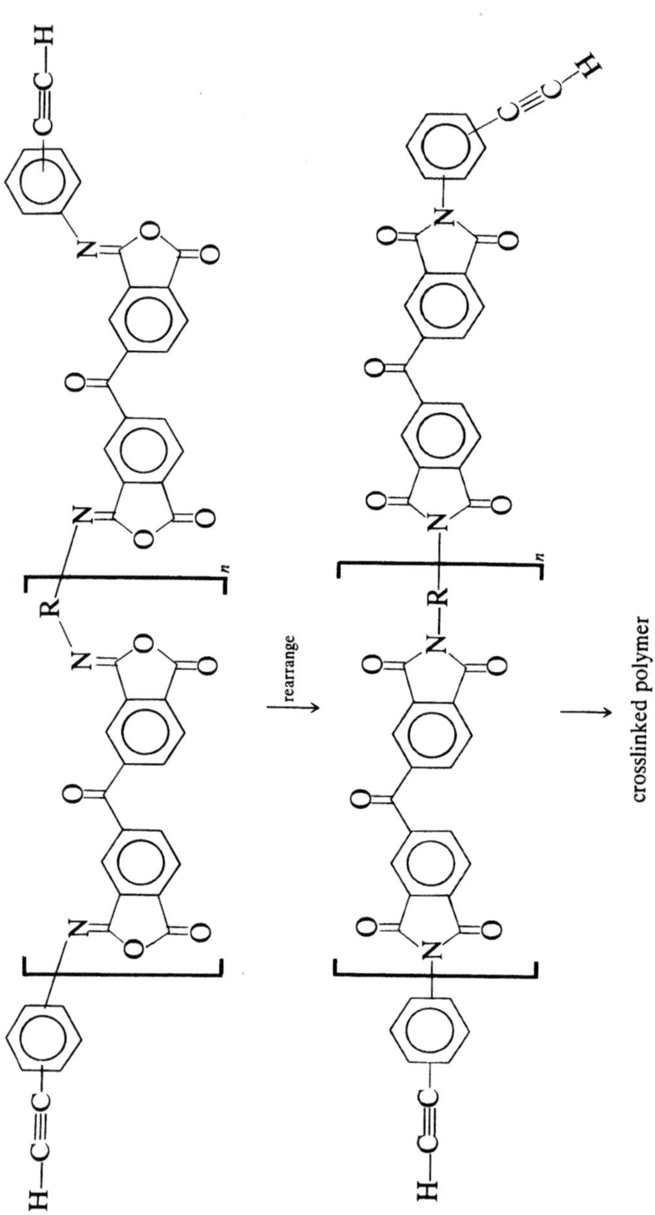

FIG. 2.5. Typical chemistry of acetylene terminated isoimides.

polyimides in conventional autoclave processing as described above for epoxy technology. However, composite parts have been successfully made using these kinds of resins,[19–25] albeit mainly for demonstration purposes.

In contrast to the PMR, LARC, P13N, etc., polyimides described previously, formulated bismaleimide resins are presently closest in processing characteristics to conventional 350°F (177°C) cured epoxies. Hence, the exceptional interest that has developed in these materials in the last few years.

2.2.3. Chemistry of Bismaleimides

Figure 2.6 shows a typical synthesis route for bismaleimides. In the first step, carried out in solvents such as *N*-methylpyrrolidone, dimethyl formamide, acetone or chloroform, at 32–77°F (0–25°C), a bismaleiamic acid is made (Fig. 2.6, eqn (6.1)). This step is analogous to those shown previously for condensation polyimides. Formation of the bismaleimide (Fig. 2.6, eqn (6.2)) occurs via either thermal or chemical methods.[26] The latter is preferred in most references in the literature,[27] although in at least one case,[28] it was found difficult to remove solvent from the bismaleimide, so that a thermal cyclisation was preferred. The complete synthesis of bismaleimides does not require isolation of the intermediate bismaleiamic acid. For example, for synthesis involving aromatic amines in dimethylformamide, the bismaleiamic acid is cyclised *in situ*.[29]

A great number of bismaleimides have been synthesised by the method described above. Table 2.1 shows a partial list of monomeric bismaleimides together with some of their relevant physical properties. Examination of Table 2.1 shows that all the bismaleimides listed are relatively high melting solids (the lowest melting point being 158°F (70°C), while the highest melt at above 572°F (300°C)).[30,31]

When heated slightly above their melting point, the viscosity of the resulting materials tends to be very low (see Table 2.1). At least one bismaleimide based on Jeffamine ED 600 or Jeffamine ED 900 (both of which are poly(oxyethylene)diamines made by Jefferson Chemical Division of Texaco) is described as an oil at room termperature.[32]

Bismaleimides such as those shown in Table 2.1 are crosslinked by thermal or chemical means. Figure 2.7 shows two crosslinking pathways for bismaleimide polymerisation which occurs through the electron deficient (and thus highly reactive) double bonds. The thermal pathway (Fig. 2.7, eqn (7.1)) usually occurs by heating the

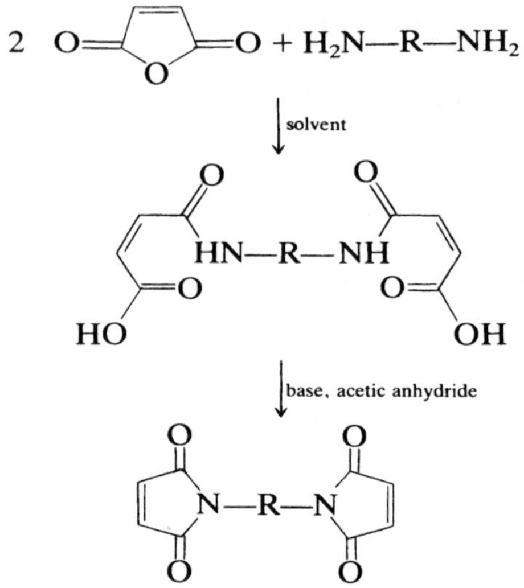

(6.1)

(6.2)

typically: R =

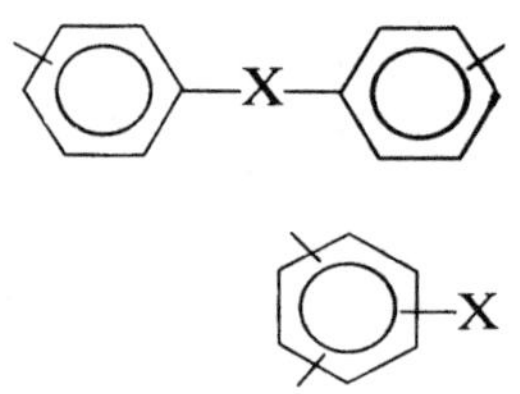

$(X = CH_2, O, SO_2, \text{etc.})$

$(X = H \text{ or } CH_3, \text{etc.})$

for rigid bismaleimides

Solvent = *N*-methylpyrrolidone, dimethylformamide, acetone, chloroform

Base = Triethylamine, sodium or potassium acetate

FIG. 2.6. Typical synthesis of bismaleimides.

bismaleimide above its melting point to 300–750°F (149–399°C), and may be catalysed by the use of peroxides and other free radical initiators,[33] as well as alkali metal salts of Bronsted acids[34] (e.g. sodium, potassium or lithium salts of acetic, succinic, maleic acids, etc., or similar salts of cyanides or carbonates). Alternative catalysts include cobalt naphthenate, dimethylaniline,[35] primary, secondary and tertiary amines or quarternary ammonium salts.[36]

The chemical crosslinking pathway is shown in Fig. 2.7, eqn (7.2) and is generally described as a nucleophilic Michael addition.[37,38] Such chemistry forms the basis of a well-known resin called Kerimid 601.

TABLE 2.1
PHYSICAL PROPERTIES OF SOME BISMALEIMIDE MONOMERS[30,31]

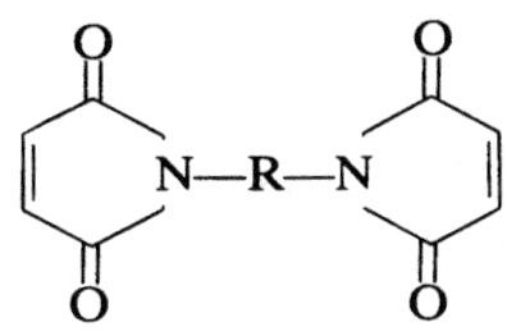

R	*Mpt (°F(°C))*	*Viscosity (cP/°F(°C))*
$—(CH_2)_2—$	374–378 (190–192)	
$—(CH_2)_6—$	279–284 (137–140)	13·2/284 (140)
$—(CH_2)_8—$	244–252 (117–122)	18·0/248 (120)
$—(CH_2)_{10}—$	232–239 (111–115)	14·5/248 (120)
$—(CH_2)_{12}—$	230–234 (110–112)	13·7/248 (120)
$—CH_2—C(CH_3)_2—CH_2—CH(CH_3)—CH_2—CH_2—$	158–266 (70–130)	15·5/266 (130)
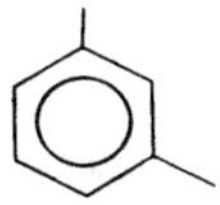	389–394 (198–200)	
	572 (300)	
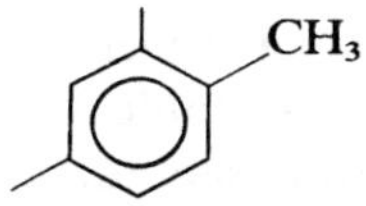	342–345 (172–174)	47·0/347 (175)
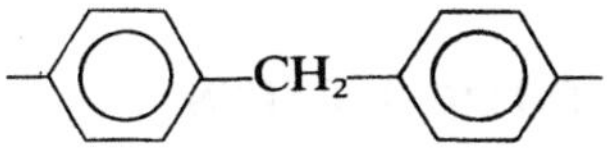	309–313 (154–156)	28·0/320 (160)
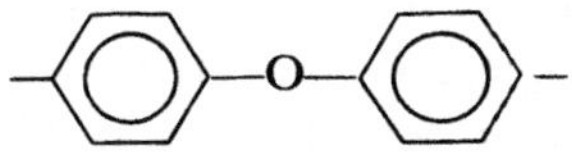	356–358 (180–181)	

Viscosity is determined by Ubbelohde method.

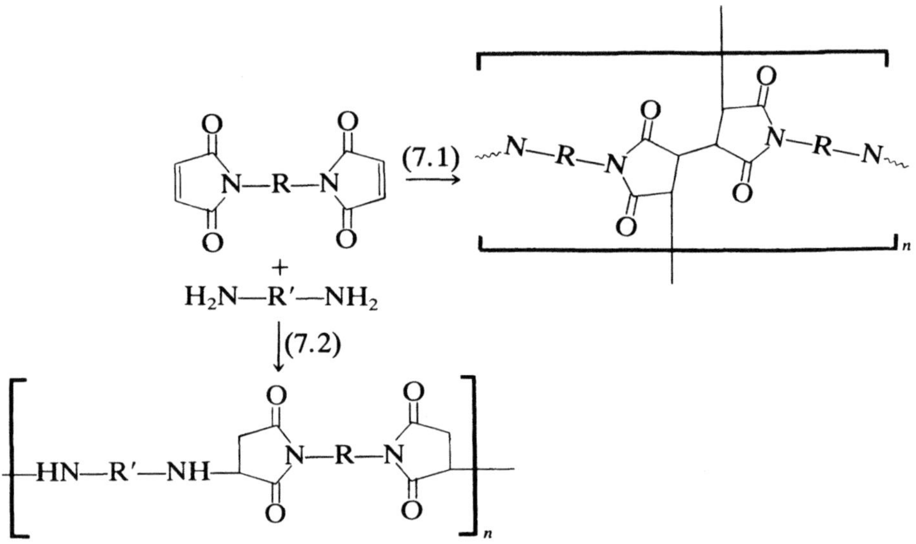

FIG. 2.7. Bismaleimide cure chemistry.

Other strong nucleophiles such as thiols and hydrogen sulphide,[39,40] as well as phenols catalysed by amines[41] may also be used to crosslink bismaleimides by routes analogous to the Michael addition shown in Fig. 2.7.

Recently, Ciba-Geigy has described[36,42] yet another interesting crosslink mechanism for bismaleimides reacted with alkenylphenols (Fig. 2.8). In this reaction pathway, the alkenylphenol first reacts with the bismaleimide via an ene reaction (Fig. 2.8, eqn (8.1)). This is followed by a Diels–Alder reaction to give a thermally stable crosslinked polymer (Fig. 2.8, eqn (8.2)). The mechanical properties of resins made by this route will be discussed later (see Section 2.3).

2.3. PROCESSABLE BISMALEIMIDE RESINS

It was shown in Table 2.1 that monomeric bismaleimides tend to be relatively high melting solids. It was also suggested that crosslinking could be done by routes which do not produce volatile by-products because of addition cure through the reactive maleimide double bonds. Bismaleimides which have viscosity characteristics like epoxy

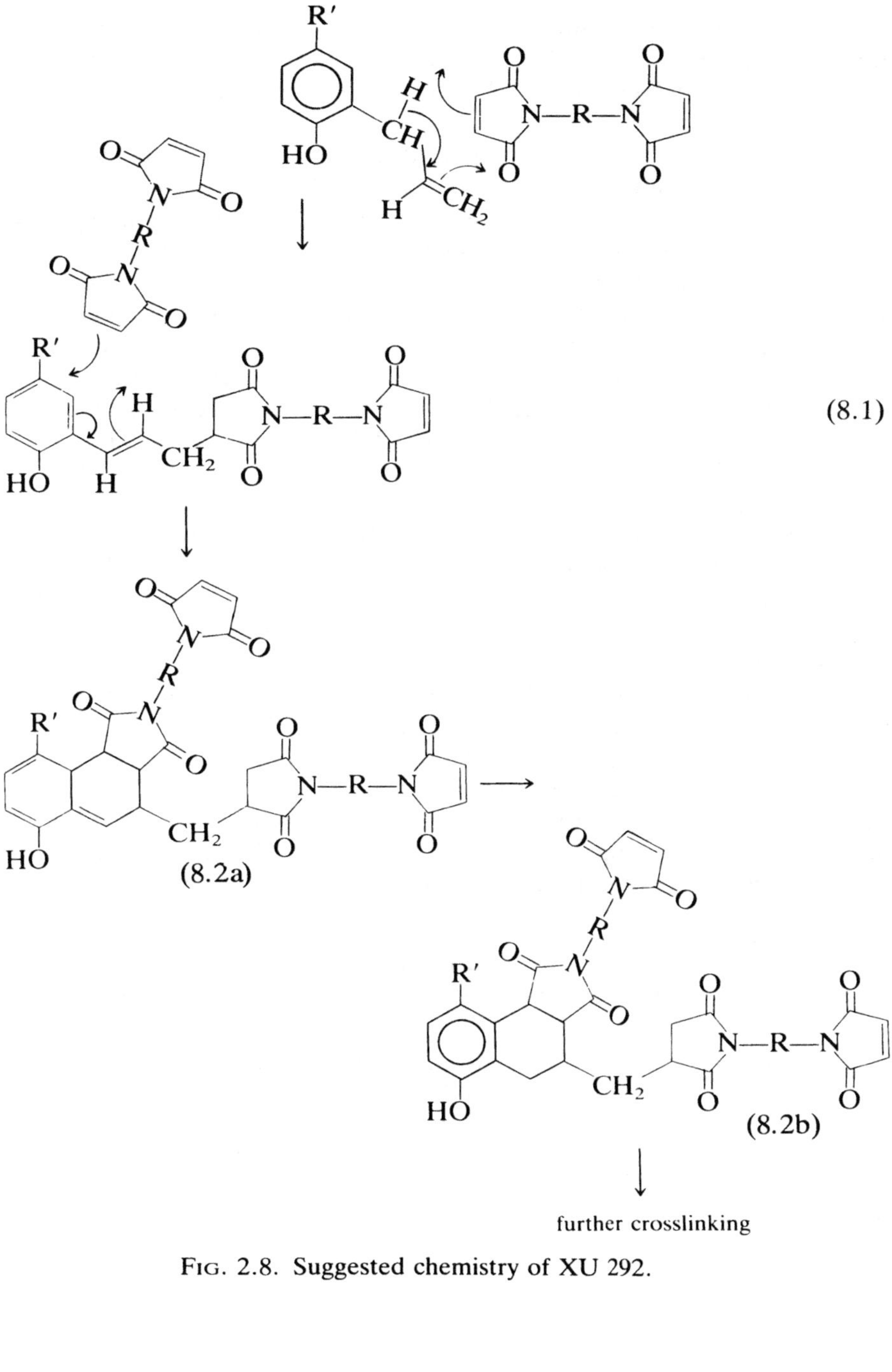

FIG. 2.8. Suggested chemistry of XU 292.

resins together with reasonable cure cycles should be processable in a manner like epoxy resins. Various methods have been used to achieve this objective, which forms the basis for the present interest in bismaleimides.

One method for achieving processable bismaleimides relates to use of the Michael addition (see Fig. 2.7). With this procedure, the reaction occurs to a sufficient stage so that crosslinking is minimised but so that the bismaleimide oligomer can be processed. Kerimid 601, made by Rhone-Poulenc,[37,38,43] is such a resin, obtained by the reaction of 4,4′-bismaleimidodiphenylmethane and methylene dianiline. This material is available as both injection and transfer moulding compounds (which have the Kinel designation) as well as a laminating resin which requires *N*-methylpyrrolidone as a processing solvent. However, the moulding temperature of 340°F (171°C) is fairly low, and good high temperature property retention at 390°F (199°C) can be achieved after a 480°F (249°C) postcure,[43,44] provided the prepreg used contains only 2% of residual solvent. Table 2.2 shows some comparative flexural strength values for Kerimid 601 E glass (Style 181 with A-1100 finish) laminates versus epoxies and silicones, which illustrates the superiority of the bismaleimide resin over other types of resins in terms of performance at elevated temperature.

Technochemie GmbH have produced several different bismaleimides which are easier to process. A low melting eutectic mixture of the bismaleimides of methylene dianiline, toluene diamine and trimethylhexamethylenediamine forms the basis of a product called H353[30] (or Kerimid 353 as marketed by Rhone-Poulenc[45]). The low melting point range of 158–257°F (70–125°C) allows this resin to be used in a hot melt process for impregnating fabrics and tows to

TABLE 2.2
COMPARISON OF FLEXURAL STRENGTH OF STYLE 108E GLASS PLAIN WOVEN LAMINATES[43,44]

Resin	*Flexural strength (Ksi) at temperatures shown*		
	77°F (25°C)	*400°F (204°C)*	*480°F (249°C)*
Kerimid 601	70	60	50
Epoxy novolac	70	32	20
Bisphenol A epoxy	80	20	15
Silicone	37	20	18

make tapes. As might be expected, this leads to prepregs with no tack or drape at room temperature, but with useful handling properties at 122°F (50°C). Typical curing conditions include: heat up to approximately 355°F (179°C) with a 1 h hold, followed by application of 50 psi (3·5 bar) pressure and further hold at this temperature for 30 min. This is followed by application of 100 psi (7 bar) pressure and holding the temperature for 4 h at 410°F (210°C). Full properties are obtained by post-cure at approximately 465°F (241°C) for 15 h. Table 2.3 shows some typical properties for prepregs of H353 using graphite, glass and aramid fibres in unidirectional lay-up. As may be seen, this resin system provides good mechanical properties through to 480°F (249°C). In addition, long term (1000 h) creep behaviour under flexural load (50% of room temperature load) up to 392°F (200°C) showed less than 0·05% creep.

Another bismaleimide produced by Technochemie is called M751[46,47] and consists of a mixture of mono and bismaleimide oligomers as shown in Fig. 2.9. When post-cured for 15 h at 480°F

TABLE 2.3
SOME TYPICAL MECHANICAL PROPERTIES FOR H353 UNIDIRECTIONAL PREPREGS[44,45]

Property	*Material*			
	Thornel 50S	*Modmor II*	*E-Glass*	*Kevlar® 49*
Fibre volume (%)	60	60	60	60
Void content (%)	1	<2	<2	<2
Density (g/cm^3)	1·51	1·54–1·60	2·1	1·39
Tensile strength (Ksi)				
room temperature	122	160·9	153·6	175·5
480°F (249°C)	119	153·6	149·2	109·7
Tensile modulus (Msi)				
room temperature	25·7	24·9	6·0	9·1
480°F (249°C)	25·2	23·7	5·6	7·3
Flexural strength (Ksi)				
room temperature	123	112·1	176·2	94·3
480°F (249°C)	109	168·2	121·4	45·3
Flexural modulus (Msi)				
room temperature	25·5	24·0	6·1	9·5
480°F (249°C)	22·6	24·7	5·8	7·2
Short beam shear (Ksi)				
room temperature	6·25	11·0	11·0	4·0
480°F (249°C)	4·7	7·2	7·2	3·0

® Registered trademark of E. I. Du Pont and Co.

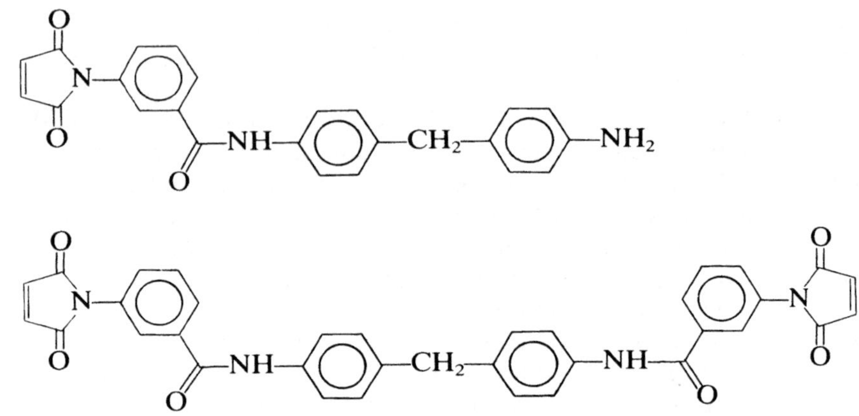

FIG. 2.9. Mono and bismaleimides in M751.

(249°C), the cured polymer has a glass transition temperature approaching 465°F (241°C) and only loses about half of its flexural strength and modulus in going from room temperature to 480°F (249°C). Unfortunately, M751 requires *N*-methylpyrrolidone as a solvent for processing.

To alleviate this problem Technochemie has produced another bismaleimide resin, called H795, which consists of an oligomer made by Michael addition of an amino acid hydrazide and a bismaleimide[46,48,49] as shown in Fig. 2.10. This product is soluble in common low boiling solvents such as toluene, 2-ethoxyethylacetate,

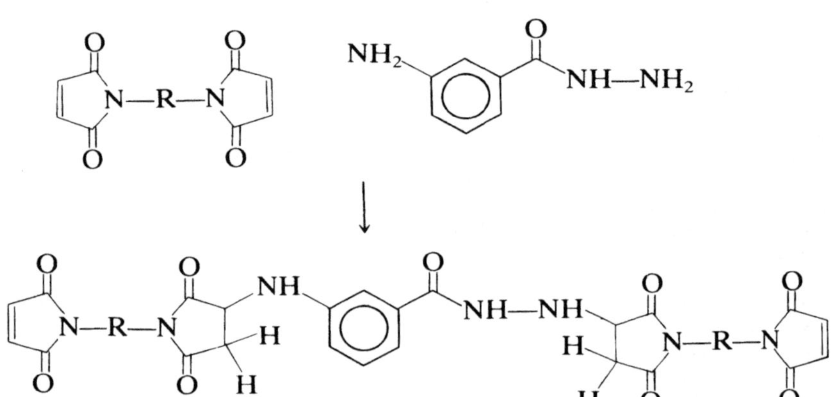

FIG. 2.10. Chemistry of H795.

diglyme, methylethylketone or methylene chloride. H795 has a relatively low melting point (at 230°F (110°C) it has a viscosity of 600–1100 cP) so it can be used for solventless hot melt impregnation of fibres. Examination of TGA data[49] for cured H795 versus M751 indicates the glass transition temperature for both resins is very similar, although based on dynamic mechanical analysis data[46] no transition is observed below 572°F (300°C). Thermogravimetric analysis indicates M751 is less thermally stable than H795. Table 2.4 compares some of the resin properties of these two resins.

It should be noted that the cure cycle for H795 used as a prepreg resin is somewhat less severe than that required for M751. For example, post-cure for 15 h at 480°F (249°C) is required for H795 versus 10 h at 392°F (200°C) and 15 h at 480°F (249°C) and 15 h at 510°F (266°C) for M751.[50]

Table 2.5 compares data for composite laminates made with M751 and H795 on Fiberite Style 134 graphite fabric. The M751 prepreg was made by solvent impregnation, whereas the H795 prepreg was made by a hot melt procedure. The large difference in retention of properties (flexural strength and modulus, as well as interlaminar shear) for H795 versus M751 is attributed to the presence of residual *N*-methylpyrrolidone which acts as a plasticiser in the M751 laminate.

Recently[46,51] Technochemie has mentioned several modifications of H795. For example, H800 (also called H795 ERK) is described as

TABLE 2.4
COMPARISON OF RESIN PROPERTIES OF M751 AND H795[46,49,50]

Property	*Test temperature (°F(°C))*	*M751*	*H795*
Density (g/cm^3)	74 (23)	1·32	1·32
Flexural strength (Ksi)	74 (23)	30·6	13·8–16·0
	482 (250)	14·8	8·7–9·0
Flexural modulus (Ksi)	74 (23)	729	725–740
	482 (250)	459	428–450
Coefficient of thermal expansion	74 (23)	17·2	20·3
(in./in. × 10^6/°F(°C))	482 (250)	46·6	40·8
Polymer decomposition temperature (°F(°C))		680 (360)	815 (435)
Melting point range (°F(°C))		230–284 (110–140)	212–266 (100–130)
Water absorption after 1000 h water immersion (%)	74 (23)	4·27	4·99

TABLE 2.5
COMPARISON OF PROPERTIES OF M751 AND H795 PREPREG LAMINATES USING FIBERITE STYLE 134 (5·5 oz/yd^2) GRAPHITE FABRIC[46,49,50]

Property	*Test temperature (°F(°C))*	*M751*	*H795*
Fibre volume (%)	74 (23)	57·9	51·0
Density (g/cm^3)	74 (23)	1·50	1·52
Flexural strength (Ksi)	74 (23)	113–131	89
	212 (100)	64	
	302 (150)	36	
	392 (200)	13	
	450 (232)	10	80
Flexural modulus (Msi)	74 (23)	11·1	11·1
	212 (100)	6·6	
	302 (150)	6·4	
	392 (200)	3·6	
	450 (232)	2·4	11·4
Interlaminar shear strength (Ksi)	74 (23)	8·8	5·3
	450	0·4	4·4

TABLE 2.6
SOME PROPERTIES OF H795E VERSUS H800 (H795ERK) UNIDIRECTIONAL GRAPHITE LAMINATES[51]

Property	*H795*		*H800 (H795ERK)*	
	68°F (20°C)	*482°F (250°C)*	*68°F (20°C)*	*482°F (250°C)*
Fibre volume (%)	60		60·8	
0° Flexural strength (Ksi)	268	180	273	181
0° Flexural modulus (Msi)	17·0	17·4	16·7	16·7
90° Flexural strength (Ksi)	7·7	4·2	11·6	5·1
90° Flexural modulus (Msi)	1·1	0·8	1·1	0·8
0° Horizontal shear (Ksi)	13·9	7·7	15·7	7·7
0 ± 45° Horizontal shear (Ksi)	7·0	6·1	8·4	6·8

Note: Prepreg tape is Celion 6000 made at 37 ± 3% resin content and areal weight 130 g/m^2 with less than 2% volatiles.

H795 modified with addition of a reactive alkyl diluent, a reactive elastomer and a small amount of catalyst such as Diazabicyclooctane (DABCO). The addition of elastomer alleviates problems with microcracking in H795. The reactive diluent is necessary to reduce the melt viscosity and improve wetability of the fibres during impregnation, as well as to improve on drape and tack. In addition, the reactive diluents enhance the thermal–oxidative stability of H800. The autoclave cure temperature was reduced from 400°F (204°C) to 335–375°F (168–191°C) by adding 0·15 to 0·25% of DABCO as a catalyst. Subsequently, it was possible to make a prepreg by hot melt processing at 158°F (70°C). Table 2.6 shows some properties of H795E (elastomer modified H795) versus H800 graphite laminates. The laminates were made from Celion 6000 graphite tape (with 1·1 to 1·3% by weight standard epoxy sizing). Considerable data on thermal ageing at 480°F (249°C), as well as hydrothermal ageing at 158°F (70°C) and 94% relative humidity for Celion 6000 laminates made with H800 are available in Ref. 51. Basically, the thermal ageing data show that H800 performs very well out to 1000 h with improvement in most properties. On the other hand the hydrothermal ageing data suggest a loss of approximately 30% in resin dominated strength properties when tested at 300°F (149°C), although flexural stiffness appears to be maintained.

A different approach to processable bismaleimide resins available for further modification by formulation is that due to Mitsubishi Gas Chemical Co., Inc. Their products are denoted 'BT' resins which refer to 'bismaleimide–triazine'. Figure 2.11 illustrates the chemistry involved in an idealised form.[52] This leads to a variety of oligomeric resins, which can be liquid or solid depending on the stoichiometry of the isocyanate monomer versus bismaleimide. The resulting oligomers have reactive end groups, characteristic of the reactive triple bonds of cyanates as well as the reactive double bonds of bismaleimides. Amongst the desirable features of these resins is their solubility in many solvents such as ketones, *N*-methylpyrrolidone, dioxane, etc., as well as compatibility with epoxy resins, acrylic acid esters and other diluents, which allows them to be readily used as impregnating resins. BT resins can be cured both with and without catalysts. The catalysts which are recommended include organometallic salts such as zinc octoate and cobalt naphthenate, tertiary amines such as triethylenediamine and *N,N*-dimethylbenzylamine, as well as organic peroxides Lewis acid salts and so on.

As expected, BT resins are sensitive to moisture in the uncured

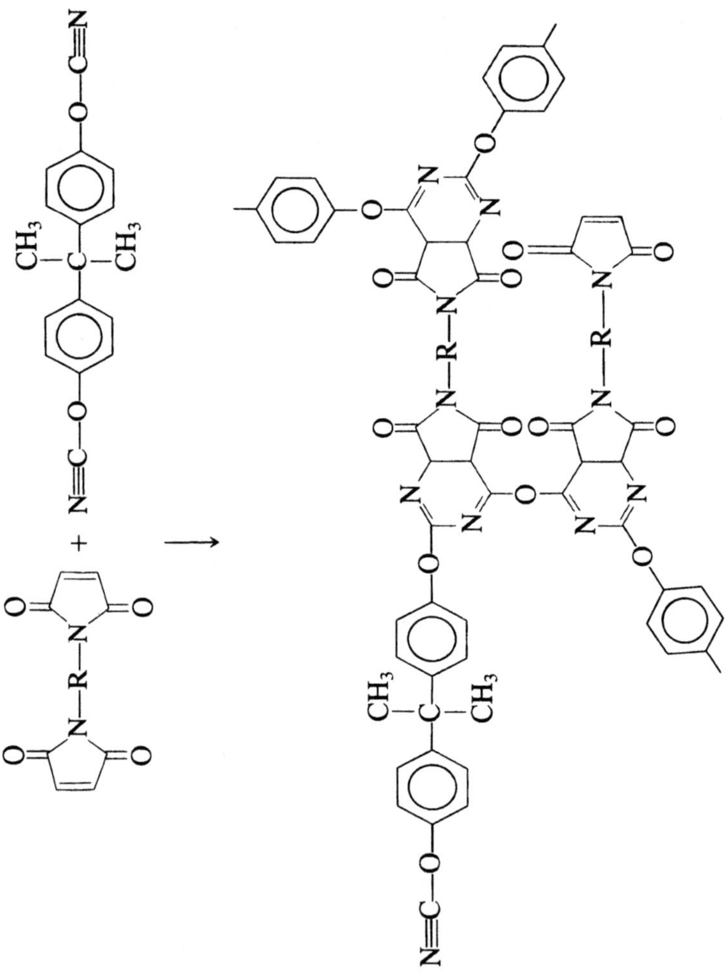

FIG. 2.11. Idealised chemistry of BT resins.

TABLE 2.7
COMPARISON OF PROPERTIES OF EPOXY AND PHENOLIC GLASS LAMINATES WITH BT2670 RESIN LAMINATE

Property	*Epoxy*	*Phenolic*	*BT2670*
Glass transition (°F(°C))	230 (110)	437 (225)	590 (310)
Resin content (%)	60	52	55
Resin flow (%)	30	25	25
Flexural strength (Ksi) at room temperature	85	64	85
Peel strength (lb/in.)			
Room temperature	9·5–10·6	6·7–7·3	9·5–10·6
302°F (150°C)	3·9–4·4	4·4–5·0	8·9–10·1

state, which requires special handling. It is claimed that they have low toxicity and odour. The many forms in which BT resins are available for purposes as wide ranging as resins for making laminates to moulding powders makes it difficult to show their potential in a short monograph. Table 2.7 compares some data for one BT resin (called BT2670) with figures for various resins such as epoxy and phenolics used in impregnating glass fabric suitable for making circuit boards. Many applications have been reported in the last few years which describe BT resins as a component of formulations. These reports mostly relate to circuit board applications, but in at least one case it is thought that BT resin is one of the components of an aerospace composite resin.

As noted in Section 2.2, Ciba-Geigy offers a very interesting bismaleimide resin system which is thought to cure by both an ene and Diels–Alder reaction. The commercial designation for this two part resin is XU 292 and consists of 4,4′-bismaleimidodiphenylmethane (Part A) as well as o,o′-diallylbisphenol A (Part B). The ease of use of this resin system derives from the solubility of the high melting bismaleimide in the liquid o,o′-diallylphenol. Table 2.8 shows typical properties[42] for the neat XU 292 resin combined at 1:1 and also 1:0·87 molar ratio of bismaleimide to o,o′-diallylphenol based on the reaction mechanism shown in Fig. 2.8. For comparison purposes, a 350°F (177°C) epoxy resin based on tetraglycidylmethylenedianiline and diaminodiphenylsulphone (TGMDA/DDS) is also shown in Table 2.8. The most interesting observations from this comparison are the higher glass transition temperatures of the XU 292 systems (by 60–90°F (16–32°C)) and the improved toughness of the bismaleimide

TABLE 2.8
COMPARISON OF XU292 BISMALEIMIDE WITH TGMDA/DDS EPOXY

Property	*XU292*		*TGMDA/DDS epoxy*
	1·0:1·0 M	*1·0:0·87 M*	
Room temperature tensile			
strength (Ksi)	11·9	13·6	8·5
modulus (Ksi)	620	564	542
elongation (%)	2·3	3·0	1·8
Room temperature flexural			
strength (Ksi)	24·2	26·8	13·3
modulus (Ksi)	589	580	499
Room temperature compression			
yield strength (Ksi)	29·9	30·4	29·2
modulus (Ksi)	348	360	284
300°F (149°C) Tensile			
strength (Ksi)	7·4	10·1	6·5
modulus (Ksi)	354	412	378
elongation (%)	2·6	3·05	1·9
HDT (°F(°C))	523 (272)	545 (284)	460 (238)
Glass transition by TMA (°F(°C))	523 (272)	540 (282)	350 (177)
Fracture toughness by compact tension G_{IC} (in. lb/in.^{2})	0·97	1·2	0·3

resin as measured by compact tension specimen testing. Other obvious features relate to improved strength retention when going from room temperature to 300°F (149°C), especially in the case of XU 292 formulated at 1:0·87 molar ratio. An outstanding feature of XU 292 appears to be the low moisture pickup (1·4–1·5%) after 2 weeks at 100°F and 100% relative humidity. Typically the TGMDA/DDS epoxy resin would pick up 4–6% moisture[53,54] under similar conditions.

In addition to the neat resin properties, some composite data[42] are also available for XU 292. This resin was used to make a tape with Hercules AS-4, 12K graphite fibre at 65% fibre volume. The tape was laid up as a 0° laminate and press cured at 350°F (177°C) for 10 to 15 min under contact pressure, followed by 1 h at 100 psi (7 bar) pressure. Post-curing was done in an air circulating oven for 2 h at 392°F (200°C) and 6 h at 482°F (250°C). Table 2.9 shows some of the laminate properties for these composites which are compared with a typical epoxy resin (AS4/3501-5A by Hercules[55]). The XU 292 composite appears to have excellent retention of short beam shear

TABLE 2.9
COMPARISON OF COMPOSITE PROPERTIES OF UNIDIRECTIONAL GRAPHITE LAMINATES MADE FROM CIBA-GEIGY XU292 BISMALEIMIDE RESIN AND HERCULES 3501-5A EPOXY RESIN

Property	*XU292*[a]		*3501-5A epoxy*[b]
	1·0:1·0 M	*1·0:0·87 M*	
Short beam shear			
Strength (Ksi)			
77°F (25°C)	16·4	17·8	18·5
350°F (177°C)	11·0	11·9	9·5
450°F (232°C)	8·6	11·4	
350°F (177°C), wet[c]	7·5	7·7	
77°F (25°C), aged[d]		15·2	
350°F (177°C), aged[d]		8·1	
Flexural strength (Ksi)			
77°F (25°C)		270	250
350°F (177°C)		219	175
350°F (177°C), wet[c]		162·5	
Flexural modulus (Ksi)			
77°F (25°C)		21·0	19·5
350°F (177°C)		21·0	18·5
350°F (177°C), wet[c]		20·8	

[a] Reinforcement is Hercules Magnamite AS-4-12K graphite fibre at 65% by volume.
[b] Reinforcement is Hercules Magnamite AS-4-12K graphite fibre at 62% by volume. Cure cycle is: Pull 20 in. (500 mm) vacuum minimum in an autoclave and heat to 225°F (107°C) in 45–75 min. Pressurise to 85 psi (6 bar) and hold at 225°F (107°C) for 1 h. Raise temperature to 350°F (177°C) in 45–75 min and hold for 1 h at 350°F (177°C).
[c] Wet = 2 weeks at 160°F (71°C) and 95% relative humidity.
[d] Aged = 1000 h at 450°F (232°C).

strength after thermal ageing at 450°F (232°C) for 1000 h (85% at 77°F (25°C); 68% at 350°F (177°C)). In addition, the 350°F (177°C) wet short beam shear and flexural strength retention appears to be quite good (65% for short beam shear; 74% for flexural). By contrast, Hercules 3501-5 resin with AS fibre in a unidirectional graphite composite drops to 73% of short beam shear strength at 300°F (149°C) and 30% of flexural strength at 300°F (149°C) after moisture conditioning.[53] Obviously, the epoxy graphite composite would have considerably lower strength after thermal ageing at 450°F (232°C).

2.4. MODIFIED BISMALEIMIDE COMPOSITES

The strategy used for making bismaleimides processable is well illustrated by the formulated resins which are available commercially. Typically, the solid bismaleimide resin is reacted or mixed in with an appropriate reactive liquid diluent such as an epoxy resin or other double-bonded diluents such as divinylbenzene, triallylisocyanurate, etc. Obviously, there are many potential diluents. However, consideration must also be given to the compatibility of the diluent with the bismaleimide resin and the degree to which it can reduce the desirable properties of the bismaleimide. A further consideration is the aesthetic, as well as toxicological aspects of the diluent, e.g. styrene is commonly used as a reactive diluent with polyesters, but has a very objectionable odour. Formulated bismaleimide resins are proprietary compositions, so the majority of this discussion will describe the physical property data with best estimates of the chemistry involved where possible.

An interesting example of formulated bismaleimides is work done at TRW by Jones *et al.*[56] for making room temperature curing polyimides to be used as sealants for fuel tanks. This is an excellent paper which illustrates the method used for formulating a bismaleimide resin with 1,4-divinylbenzene or an epoxy resin as a diluent. The various bismaleimide components of the formulated resins consist of mixtures of bis and tris maleimides of Jeffamines and 4,4′-bismaleimidodiphenylmethane. The preferred compositions consist of Jeffamine ED900 bismaleimide, 4,4′-bismaleimidodiphenylmethane and Jeffamine AP-22 (an oligomer from the Mannich reaction of methylenedianiline and formaldehyde) together with DER 736 (a Dow Chemical aliphatic epoxy resin). Jeffamine AP-22 is a curing agent for the epoxy resin, but as noted in Section 2.2 may also react with the bismaleimide in a Michael reaction to complete crosslinking. Table 2.10 shows some results for peel adhesion strengths of the formulated bismaleimides compared with a polysulphide sealant which might normally be used in this application. The authors[56] concluded the bismaleimide sealant was superior to the polysulphide for various reasons including high adhesion under conditions of minimum surface preparation and relatively low to medium commercial cost.

Another group at TRW worked an alternative technology for making bismaleimide resins with excellent moisture resistance.[57] These resins, unlike those described by Jones *et al.*,[56] are intended for

TABLE 2.10

COMPARISON OF PEEL ADHESION STRENGTHS FOR A FLEXIBLE BISMALEIMIDE SEALANT VERSUS A POLYSULPHIDE SEALANT[56]

Formulation	*Peel strength at room temperature (lbs/in.)*	
	No fuel ageing	*After fuel ageing at 140°F (60°C) for 7 days*
Polysulphide 1422	17–25	18–23
Bismaleimide/epoxy in ratio:		
90/10	13–16	
80/20	18–45	24–39
75/25	61–64	52–63
70/30	14–47	
60/40	3–6	3–6

Notes:

(1) Substrates are uncleaned aluminium which is coated with a urethane primer corresponding to MIL-C-27725.

(2) Fuel used is Jet Reference Fuel (MIL-S-8802D). Peel strength was determined according to MIL-S-8802D.

(3) The bismaleimide/epoxy formulation is a mixture of Jeffamine ED900 bismaleimide, 4,4′-bismaleimidodiphenylmethane, Jeffamine AP-22 and DER736 epoxy (see text for further description).

structural applications to make fibre reinforced composites. The resins are formulated as mixtures of polybutadiene resins, 4,4′-bismaleimidodiphenylmethane, an epoxy resin (DEN 438), a peroxide and chromium octoate. Table 2.11 shows a detailed formulation, while Table 2.12 shows some graphite composite properties for this formulation. In a similar vein, Browning[58] described work done by TRW and Hercules in support of an Air Force Materials Laboratory programme for improved moisture resistant resins for composite fabrication. The formulations of the resins described by Browning[58] are somewhat similar to those described by Vaughn *et al.*,[57] i.e. they consist of mixtures of epoxy resins, polybutadiene resins, 4,4′-bismaleimidodiphenylmethane, a peroxide and an amine such as benzyl dimethylamine or diaminodiphenylsulphone. The most important feature of these resins is that they have improved moisture resistance compared to conventional epoxies and the composites are made with no-bleed vacuum bag cures. In addition these resins have good room temperature out-time and low density. Two resins viz. a 275°F (135°C) and a 350°F (177°C) service resin were developed, but no commercial products have appeared as a result of this technology.

TABLE 2.11
FORMULATION OF POLYBUTADIENE BISMALEIMIDE RESIN AT TWR[56]

Ingredient	*Percent by weight*
HYSTYL C2000 carboxy terminated 1,2-polybutadiene[a,c]	24·2
G2000/SA adduct of hydroxy terminated 1,2-polybutadiene, 2000 MW and succinic anhydride (3:2 mole ratio)[b,c]	48·4
4,4′-bismaleimidodiphenylmethane	18·1
DEN438 epoxy novolac resin	4·5
Lupersol 101 organic peroxide[d]	4·5
Chromium octoate	0·3

[a] Available from Nippon Soda Co., Ltd as Nisso PB B2000.
[b] Available from Nippon Soda Co., Ltd as Nisso PB GM2000.
[c] HYSTL C2000 and G2000/SA were products previously made in the US by TRW.
[d] Available from Pennwalt Corp.

A commercially available bismaleimide resin is marketed by Hexcel as a product called F178. Some insight into the formulation of this resin comes from an article published by Nash *et al.*[59] in which analysis was performed on F178. Amongst the ingredients are: 4,4′-bismaleimidodiphenylmethane, triallylisocyanurate and probably a free-radical producing catalyst. The article by Nash *et al.*[59] also gives some insight into potential uses for bismaleimide resins, viz. for operational requirements at temperatures in excess of 400°F (204°C) in future vertical and short-take-off and landing aircraft (VSTOL) where large and heavy engines are required for vertical lift. Poveromo[60] has also mentioned the use of F178 in applications at Grumman where thermal envelopes between 260–450°F (127–232°C) are required, together with greater thermal oxidative stability and resistance to moisture degradation than epoxies. The location of use of F178 on graphite or glass fibre is stated[60] as being on the DC-8 Nacelle Deicer Plenum. Initial applications at Grumman occurred in the mid 1970's in which bismaleimide fibreglass radomes were constructed. Apart from the thermal capabilities of F178, processability like that for state-of-the-art epoxies without solvents emitted during cure, were important factors in the use of this bismaleimide prepreg, as well as another product made by US Polymeric called V378A. An extensive data base[61] has been gathered on F178 resin impregnated onto graphite and

TABLE 2.12
PROPERTIES OF POLYBUTADIENE BISMALEIMIDE GRAPHITE[a] UNIDIRECTIONAL COMPOSITES (see Table 2.11 for formulation)

Property	*Value*
Uncured tape:	
% volatiles by weight	0·8
% resin content by weight	29·3
Cured laminate:	
% resin content by weight	27·4
density (g/cm^3)	1·5
fibre volume (%)	62
Voids (%)	<1
Flexural strength (Ksi)	
75°F (24°C)	214
350°F (177°C)	182
350°F (177°C), wet[b]	173
Short beam shear strength (Ksi)	
75°F (24°C)	7·8
350°F (177°C)	5·0
350°F (177°C), wet[b]	4·7

[a] Graphite fibre is Hercules Magnamite AS. Cure cycle for making the laminate is claimed to be approximately 350°F (177°C) using conventional vacuum bag (i.e. 15 psi (1 bar)) or autoclave moulding techniques.
[b] Wet is not defined but assumed to be moisture saturation at 160°F (71°C) and 100% relative humidity.

glass substrates in a hybridised polymer matrix composite. Table 2.13 shows a comparison of composite properties made for an all epoxy (350°F (177°C) system) glass–graphite hybrid versus an equivalent F178/glass–graphite hybrid. The construction of this hybrid is quite complex and is also shown in Table 2.13 (see footnotes). From this data it is quite apparent that at least below 260°F (127°C), the epoxy system is as good as, if not superior to the F178 bismaleimide system. However, F178 has a higher glass transition temperature (491°F (255°C)) than the epoxy (414°F (212°C)). The cure cycles for both systems are comparable viz. autoclave cure of 1 h at 350°F (177°C) at 85 psi (6 bar) and oven postcure of 4 h at 350°F (177°C). A higher temperature postcure at (480°F (249°C)) would tend to improve F178 properties.[59] A thinner 10 ply laminate of F178 glass–graphite

TABLE 2.13

COMPARISON OF PHYSICAL PROPERTIES FOR A STYLE 7781 WOVEN GLASS–CELANESE CELION 6000 UNIDIRECTIONAL GRAPHITE/F178[a] BISMALEIMIDE HYBRID[d] COMPOSITE VERSUS A STYLE 7781 WOVEN GLASS/F166[a] EPOXY–HERCULES AS-1 UNIDIRECTIONAL GRAPHITE/3501-5A[b] EPOXY HYBRID[d] COMPOSITE[61]

Property	*Bismaleimide hybrid*	*Epoxy hybrid*
% Resin content by weight	46·1	42·3
% Fibre content by volume	47·0	48·9
% Voids by volume	negligible	negligible
Specific gravity (g/cm^3)	1·63	1·61
Flexural strength (Ksi)		
75°F (24°C)	110·0	155·7
260°F (127°C)	111·9	140·2
260°F (127°C), wet[c]	131·0	129·8
Flexural modulus (Ksi)		
75°F (24°C)	787	745
260°F (127°C)	764	737
260°F (127°C), wet[c]	860	797
Interlaminar shear strength (Ksi)		
75°F (24°C)	3·88	6·27
260°F (127°C)	3·87	5·76
260°F (127°C), wet[c]	3·67	5·64
Glass transition (°F(°C))	491 (255)	414 (212)
Glass transition, wet[c] (°F(°C))	220 (104)	196 (91)

[a] F178 is a Hexcel bismaleimide resin and F166 is a Hexcel 350°F (177°C) epoxy resin.
[b] 3501-5A is a Hercules 350°F (177°C) epoxy resin.
[c] Wet = 21 days at 140°F (60°C) and 98% relative humidity.
[d] Laminates were autoclave cured for 1 h at 350°F (177°C) and 85 psi (6 bar) pressure and post-cured in an oven for 4 h at 350°F (177°C). Laminate construction is: (45°WG, 135°WG, $(90°G)_2$, 135°G, 45°G, $(0°G)_7$, 45°G, 135°G, $(0°G)_7$, 135°G, $45°G)_S$
where
WG = 7781 woven glass
G = graphite fabric
S = symmetric laminate consisting of 50 plies total
and 0°, 45°, 90°, 135° refer to the fibre directions.

delaminated[61] and illustrates one of the problems of this resin, viz. a tendency to microcrack.

As mentioned previously, a US Polymeric bismaleimide prepreg resin (V378A), was also considered by Grumman.[60] According to the literature[62] this resin shows none of the microcracking problems associated with F178. The reason for this behaviour is claimed[63] to be due to the use of mixtures of divinyl benzenes which gel the bismaleimide at low temperatures. The addition of small amounts

(0–15%) of compatible elastomers such as polyetherpolysulphones with molecular weights 10 000 to 20 000, linear homopolymers of bisphenol epoxies with molecular weights from 40 000 to 12 000 and/or polyacrylic esters like ethylhexylacrylate also contribute to transverse strength of uncured and cured unidirectional tape high modulus graphite composites. The majority of V378A consists of a mixture of bismaleimide resins whose composition resembles H353.[30] Since divinylbenzenes are liquids, these materials help make V378A processable by allowing formulating freedom. However, aesthetically, the divinylbenzene additives detract from the resin's properties because of their volatility (V378A may appear dry depending on how carefully it is stored) and distinctive odour.

Despite this drawback V378A is probably the most successful bismaleimide composite matrix resin in today's market. The reason for this probably lies in the close relationship developed by US Polymeric and General Dynamics in practical applications. In addition, the data base for V378A composites is probably the largest one available[64–66] for a bismaleimide.

Amongst the applications at General Dynamics, V378A prepregs are being used to make complete wing skins and ribs for the F16XL 'cranked arrow' fighter plane.[64] V378A is also being evaluated for construction of a wing flap in another prototype aircraft and in the firewall of an advanced concept helicopter.[64] The latter application stems from the low-smoke density and flammability which is typical of bismaleimide resins.

Table 2.14 shows some of the neat resin properties of V378A. Table

TABLE 2.14
RESIN PROPERTIES OF V378A BISMALEIMIDE[62,65]

Property	*Value*
Tensile strength (Ksi)	
75°F (24°C)	11·3
450°F (232°C)	6·4
Elongation (%)	
75°F (24°C)	6·6
450°F (232°C)	6·5
Specific gravity (g/cm^3)	1·27
Viscosity (cps)	
100°F (38°C)	*ca.* 50 000
200°F (93°C)	*ca.* 8 000

TABLE 2.15
SOME MECHANICAL PROPERTIES OF V378A/T300-6K UNION CARBIDE UNIDIRECTIONAL GRAPHITE COMPOSITES[64,66]

Property	*Initial*	*Humidity aged*[a]
0° Flexural strength (Ksi)		
75°F (24°C)	265·0	254·0
350°F (177°C)	197·5	135·0
450°F (232°C)	179·0	
550°F (288°C)	122·0	
600°F (316°C)	107·0	
0° Flexural modulus (Msi)		
75°F (24°C)	19·8	18·4
350°F (177°C)	20·7	15·3
450°F (232°C)	19·2	
550°F (288°C)	18·4	
600°F (316°C)	17·6	
0° Horizontal shear (Ksi)		
75°F (24°C)	18·3	13·7
350°F (177°C)	10·9	7·4
450°F (232°C)	9·2	
550°F (288°C)	6·4	
600°F (316°C)	5·8	
90° Tensile strength (Ksi)		
75°F (24°C)	9·0	4·6
350°F (177°C)	5·9	1·6
90° Tensile strain (%)		
75°F (24°C)	0·66	0·31
350°F (177°C)	0·59	0·18
Specific gravity (g/cm^3)	1·60	
Void content (%)	negligible	
Fibre volume (%)	64·8	
% Resin content by weight	28·6	

[a] Humidity aged = 8 weeks at 160°F (71°C) and 98% relative humidity. See text for cure conditions.

2.15 shows some moisture effects on V378A unidirectional graphite composites which illustrates one of the major advantages of bis-maleimides over epoxy resins, viz. serviceability at temperatures of 350°F (177°C) in high humidity environments. Table 2.16 shows some thermal ageing data for periods up to 9 months at 350°F to 450°F (177° to 232°C) for a unidirectional graphite composite made with V378A. The supporting data base for V378A composites is very large, so the reader is referred to the references[64,66] for further details.

TABLE 2.16
SOME MECHANICAL PROPERTIES OF THERMALLY AGED V378A/T300 UC309 UNIDIRECTIONAL GRAPHITE COMPOSITES[64,66]

Property	*Initial*	*Aged 6 months*[a]	*Aged 9 months*[b]
0° Flexural modulus (Msi)			
75°F (24°C)	298	320	267
350°F (177°C)	245	271	
450°F (232°C)	233		205
0° Short beam shear strength (Ksi)			
75°F (24°C)	15·8		15·5
350°F (177°C)	10·1		
450°F (232°C)	7·7		8·6
% Weight loss by weight		0·6	3·4
% Resin content by weight	27–29		
% Fibre volume	64–66		
% Void content	0–2·6		

[a] Thermally aged at 350°F (177°C) for 6 months.
[b] Thermally aged at 450°F (232°C) for 9 months.

In addition to the mechanical properties, several references[62,67] are useful in discussing the processing technology associated with V378A. The resin apparently processes similarly to conventional epoxies. The prepregs typically are no-bleed systems which make net resin composites. When they are fresh V378A prepregs have reasonably good tack which allows repositioning of the prepreg. However, after being open at ambient for several hours, considerable tack is lost, probably because of the evaporation of the divinyl benzenes. A typical cure cycle for V378A prepregs involves application of 85 psi (6 bar) in an autoclave prior to start of heating and throughout the cure cycle. The material is heated at approximately 3°F (1·6°C) per minute to a dwell temperature of 220°F (104°C) and is held at this temperature for 1 h during which compaction and ply knitting can occur because of the low viscosity of the resin. Heating up to 350°F (177°C) with a hold of 4 h at this temperature is followed by a post-cure step which is done in an oven without restraint for 4 h at 475°F (246°C). This kind of processing is well within the reach of present manufacturing technology in the aerospace industry.

Another formulated bismaleimide resin for composites is Rigidite 5245C made by Narmco Materials, Inc. Unlike F178 and V378A,

Rigidite 5245C is claimed as a high impact resin with properties which might make it suitable for construction of primary structure components of aircraft.[68] Once again, the processing of this resin is epoxy-like, with no postcure being required. A typical cure schedule is to apply 45 to 100 psi (3 to 7 bar) autoclave pressure and vent the bag at 25 psi (1·7 bar) before applying heat. The lay-up is then heated up at 1 to 3°F (0·5 to 1·6°C) per minute to 350°F (177°C) and held at this temperature for 2 h.

The chemistry of Rigidite 5245C is not published, but it is thought to involve the use of 'BT' resins (see Section 2.3). Examination of the rheometrics curve for Rigidite 5245C shows two minima[68,69] which suggests two separate resin components with one component melting into the second. In all likelihood, given what is known about 'BT' resins, the other component is an epoxy which provides the fluidity necessary for formulating.

Table 2.17 shows some properties of a unidirectional graphite

TABLE 2.17

SOME MECHANICAL PROPERTIES OF RIGIDITE 5245C UNIDIRECTIONAL GRAPHITE (WITH 1.4% STRAIN FIBRE) COMPOSITE VERSUS NARMCO EPOXIES 5208 AND 5241 MADE AS COMPOSITES WITH THE SAME FIBRE[68]

Property	*Rigidite 5245C*	*5208*	*5241*
0° Compressive strength (Ksi)			
75°F (24°C)	240	207	200
200°F (93°C)	220	195	175
270°F (132°C)	175	170	
270°F (132°C), wet[a]	165	140	
0° Short beam shear strength (Ksi)			
75°F (24°C)	16	18	16
200°F (93°C)	15	15	13
270°F (132°C)	13	12	6
270°F (132°C), wet[a]	9	8	3
Instrumented impact[b]			
P_f (lbs)	460	125	240

[a] Wet = 40 h water boil prior to test and then hold specimen for 5 min at test temperature before testing.

[b] Lay up for instrumented impact test is a 12 ply quasi-isotropic laminate. The test equipment is a Dynatup instrument which is essentially a Gardener impact device and records all features of load penetration as a function of time.

(normal strain fibre—1·4%) composite made with Rigidite 5245C versus a Narmco 5208 (350°F (177°C) epoxy) resin and Narmco 5241 (a 200°F (93°C) somewhat tough epoxy resin). The P_f (instrumented impact) value represents the maximum load absorbed during a low velocity impact on a 12 ply quasi-isotropic laminate. From this data, it is clear that Rigidite 5245C is relatively tougher than a so-called 'toughened' epoxy. In addition, the wet strength retention (after 40 h water boil) of this composite is better than that for an epoxy (although apparently worse than V378A—see Table 2.15 short beam shear strength values at 350°F (177°C)). This probably reflects a lower glass transition temperature for Rigidite 5245C. The reader is referred to Refs 68 and 69 for more detailed data in using Rigidite 5245C with high strain graphite fibres and in particular results for open hole tension and compression tests, as well as compressive strength after impact.

As a result of the objectionable odour and limited useful tack life of V378A, there has been much interest in formulated bismaleimides which alleviate these problems. One resin which is claimed[70] to be useful is made by Ciba-Geigy (Fountain Valley, California) and is called R6451. The ingredients used to make up R6451 are proprietary, but probably do not consist of Ciba-Geigy's XU 292 system if the no-odour claim is correct. However, there appears to be about 6 to 8% of a reactive diluent in a 32 to 44% resin content prepreg of R6451.

Table 2.18 shows some typical laminate properties of R6451 on T300-6K unidirectional graphite fibre. The cure cycle is similar to that for V378A, viz. a hold at 175°F (79°C) for 2 h, a hold at 350°F (177°C) for 4 h (all done at 85 psi (6 bar) pressure in an autoclave) followed by an oven post-cure at 475°F (246°C) for 4 h. The hold at 175°F (79°C) has been found to be unnecessary to give equivalent properties as shown in Table 2.18. However, cure of 1 h at 280°F (138°C) and 1 h at 350°F (177°C) with 85 psi (6 bar) pressure, followed by 4 h at 425°F (218°C) post-cure leads to a deterioration in 350°F (177°C), wet short beam shear and flexural strengths.

R6451 unidirectional graphite fibre laminates were thermally exposed at 350°F (177°C) for lengthy times (up to 2000 h), and showed good flexural and short beam shear strength retention at ambient as well as 350°F (177°C). However, up to 6·3% weight loss was experienced after 2000 h at 450°F (232°C). A quasi-isotropic R6451 graphite fibre laminate which was heat aged for 25 h at 350°F (177°C)

TABLE 2.18
SOME MECHANICAL PROPERTIES OF R6451/T300-6K UNIDIRECTIONAL GRAPHITE LAMINATES[70]

Property	*Cure cycle/value*				
	A^a		B^b	C^c	
	Dry	*Wet*[d]	*Dry*	*Dry*	*Wet*[d]
0° Flexural strength (Ksi)					
75°F (24°C)	280	260	265	245	
350°F (177°C)	229	180	210		114
450°F (232°C)	189	120	190		
0° Flexural modulus (Msi)					
75°F (24°C)	18	18	20	18	
350°F (177°C)	17	17	20		14
450°F (232°C)	16	15	17		
0° Short beam shear strength (Ksi)					
75°F (24°C)	16·3	14·0	16·3	15·0	
350°F (177°C)	11·6	7·8	10·0		6·4
450°F (232°C)	9·6	6·3	7·4		

[a] Cure cycle is 2 h/275°F (135°C) + 4 h/350°F (177°C) under 85 psi (6 bar) pressure + 4 h/475°F (246°C) (unrestrained). Fibre volume is 65%.
[b] Cure cycle is 4 h/350°F (177°C) under 85 psi (6 bar) pressure + 4 h/475°F (246°C) (unrestrained). Fibre volume is 65%.
[c] Cure cycle is 1 h/280°F (138°C) + 1 h/350°F (177°C) under 85 psi (6 bar) pressure + 4 h/425°F (218°C) (unrestrained). Fibre volume is 60%.
[d] Wet = 15 days at 150°F (66°C) and 95% relative humidity.

and 264 h at 275°F (135°C) showed no evidence of microcracking when cross-sectioned and examined under a microscope.

A different approach to modified bismaleimides uses reactive oligomers[71] rather than diluents. The oligomers shown in Fig. 2.12 are vinylstilbazole (VST) and vinylpolystyrylpyridine (VPSP). When mixed with Technochemie's H795 (see Section 2.3) VST or VPSP have the effect of lowering the cure temperature of H795 by about 200°F (93°C) and shortening the cure cycle. Unfortunately, the formulated mixtures are solids at room temperature (melting point 160–212°F (71–100°C)), but they can be dissolved in tetrahydrofuran. Table 2.19 compares some mechanical properties of H795 and formulated H795/VST, H795/VPSP graphite fibre laminates. It can be seen that short beam shear and flexural strengths are improved by this formulating approach. However, moisture absorption after 2 h of water boil seems to be somewhat greater for the formulated systems compared to H795 alone. A comparable epoxy graphite fibre composite would absorb 2 to 3% water under similar conditions.

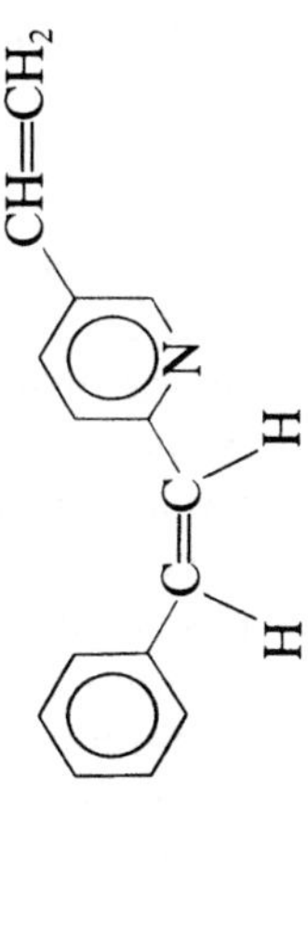

vinylstilbazole (VST)

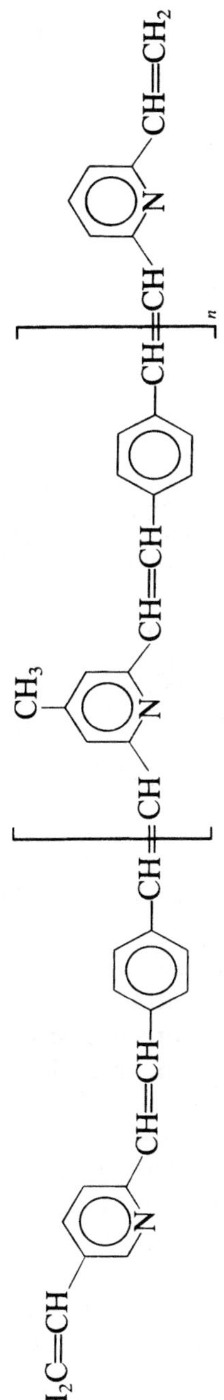

vinylpolystyrylpyridine (VPSP)

FIG. 2.12. Reactive diluents used by Parker *et al.*[71]

TABLE 2.19
SOME MECHANICAL PROPERTIES OF H795 AND H795/VST, H795/VPSP GRAPHITE COMPOSITE LAMINATES MADE WITH 9-PLY SATIN-WEAVE GRAPHITE FIBRE[71]

Property	*H795*	*H795/VST (1:4)*	*H795/VPSP (1:4)*	*VPSP*
% Resin content by weight	30·5	30·0	26·0	26·0
Density (g/cm^3)	1·55	1·39	1·43	1·44
Short beam shear strength (Ksi)				
75°F (24°C)	1·7	2·96	2·95	3·15
Flexural strength (Ksi)				
75°F (24°C)	24	46	49	49
Flexural modulus (Ksi)				
75°F (24°C)	7·7	7·3	7·9	7·6
% Water absorption by weight[a]	0·72	0·99	1·17	1·36

[a] Water absorption determined after boiling in water for 2 h.
Note: The H795/VST and H795/VPSP prepregs were press cured at 392°F (200°C)/75–100 psi (5–7 bar) pressure for 3·5 h. H795 was also press cured under these conditions, but was not fully cured as shown by dynamic mechanical analysis.[71]

Other formulated bismaleimide matrix resins for making composite laminates are in various stages of development by such companies as AVCO Corp., Hexcel (HX1516), Fiber Resin Corp. (BMS 9200) and Hysol, in the United States. The reader may refer to any of these companies for further information.

2.5. APPLICATIONS IN ADHESIVE AND ELECTRONICS FOR BISMALEIMIDES

2.5.1. Bismaleimide Adhesives

While there are many patents covering maleimide containing adhesives, very little open literature discusses these materials. Typically, the bismaleimide component is added to any one of the usual resins used in adhesive formulations such as cyanoacrylates, acrylics, isocyanates, epoxies, polyesters and even polyamides (to make a hot melt crosslinkable adhesive). The bismaleimides improve the thermal stability characteristics of the adhesive. Bonding applications include electrical circuits, rubber to glass bonding and anti-fouling of ship hulls to prevent growth of crustaceans and algae. Most patents in these areas appear to be Japanese and include companies such as Tokyo Shibaura Electric Co. Ltd, Toray Industries, Yokohama Rubber Co.

Ltd, Ihara Chemical Industry Co. Ltd, Nippon Zeon Co. Ltd and General Electric.

Obviously, any of the modified bismaleimides discussed previously for use as matrix resins in composite structures are potential adhesive candidates. However, none of these have been specifically mentioned for this use, with the exception of BT resins A110 and A120.[72] The latter resins are supplied as 65% by weight solids dissolved in methyl ethyl ketone and are described as useful for bonding copper foil to Kapton (Du Pont polyimide film). These adhesives are applied to the Kapton and flash dried for 15 to 40 min at 248–302°F (120–150°C). Then a lengthy cure cycle of 4 h at 140°F (60°C), 4 h at 176°F (80°C), 4 h at 230°F (110°C) and 6 h at 302°F (150°C) gives adhered substrates with peel strength of 12 to 15 lbs/in.^{2} (83 to 103 kN/m^2) at 77°F (25°C) and 3 to 6 lbs/in.^{2} (21 to 42 kN/m^2) at 374°F (190°C).

The most complete product line of modified bismaleimide adhesives is that of Hysol Division, The Dexter Corp. Included in this product line are film adhesives, pastes, syntactics and primers. Properties of these materials have been reported,[73,74] although the chemistry has not been revealed other than that it is 'bismaleimide/epoxy using somewhat similar addition curing methods to those reported by others' in one case, and 'a unique curing mechanism' for another product.

The products in the first case are a film adhesive carried on a glass scrim cloth (called EA9655), a paste adhesive (called EA9351) and a companion syntactic paste (called EA9367), as well as a primer (called LR100-581 in the references, but now called XEA9281 according to a recent technical bulletin). For illustrative purposes, Table 2.20 shows some adhesive properties of EA9655 when used to bond substrates including aluminium, titanium, stainless steel and polyimide/graphite fibre composites.[74] Also included in Table 2.20 is some data for an epoxy film adhesive (Hysol EA9649), from which it can be seen that for moderate temperatures (below 350°F (177°C)), initial epoxy adhesive properties are better, but at higher temperatures and also with prolonged elevated temperatures, the bismaleimide adhesive is superior. The paste adhesives have somewhat similar adhesive properties to those of EA9655.

Table 2.21 shows some adhesive properties for a film adhesive which uses 'a unique curing mechanism' (called LR100-573, but known in a more recent Hysol technical brochure as XEA9673). From the data shown in Table 2.21, it can be seen that this film adhesive appears to be usable at temperatures about 50°F (26°C) higher than EA9655. In

TABLE 2.20

ADHESIVE PROPERTIES OF HYSOL BISMALEIMIDE FILM ADHESIVE EA9655[a] COMPARED TO HYSOL EPOXY FILM ADHESIVE EA9649[b]

Test temperature (°F(°C))/substrates	*Tensile shear strength (Ksi)*	
	EA9655	*EA9649*
77/Aluminium[c]	2·7	3·3
300/Aluminium[c]		3·3
420/Aluminium[c]		1·8
450/Aluminium[c]	2·9	
500/Aluminium[c]	1·8	
550/Aluminium[c]	0·9	
Aged 18 h at 365°F (185°C)/Aluminium		2·9
Aged 500 h at 450°F (232°C)/Aluminium	2·0	
500/6A14V Titanium[d]	1·4–2·6	
400/301 Stainless steel	1·8	
77/V378A Graphite composite[e]	2·1	
350/V378A Graphite composite[e]	1·8–3·3	
500/V378A Graphite composite[e]	1·9–2·4	
77/PMR-15 Graphite composite[f]	1·8	
350/PMR-15 Graphite composite[f]	2·0	
450/PMR-15 Graphite composite[f]	1·5	

[a] EA9655 used weighs 0·06 lbs/sq. ft and is supported on a glass scrim fabric. Cure cycle is 1 h at 350°F (177°C)/25 psi (1·7 bar) pressure, followed by an unrestrained post-cure of 2 h at 450°F (232°C) for aluminium substrates, and 2 h at 475°F (246°C) for all other substrates.

[b] EA9649 used weighs 0·06 lbs/sq. ft and is unsupported. Cure cycle is 1 h at 350°F (177°C)/30 to 100 psi (7 bar) pressure with a heat up rate of 1·5 to 9°F/min (0·8 to 5°C/min).

[c] The aluminium adherends are FPL etched bare 2024-T3. In the case of EA9649, the adherends were primed with Hysol EA9205 primer before bonding.

[d] Surface preparation on titanium: degrease, alkali clean, HF/HNO_3 etch, phosphate treat, wash and dry.

[e] V378A graphite composite is a US Polymeric bismaleimide resin prepreg on Union Carbide T300/3K unidirectional graphite.

[f] PMR-15 graphite composite is a NASA, Lewis polyimide resin prepreg on Celanese Celion woven graphite fibre.

TABLE 2.21
ADHESIVE PROPERTIES OF HYSOL BISMALEIMIDE[73,74] FILM ADHESIVE XEA9673[a]

Test temperature (°F(°C))/substrate	*Tensile shear strength (Ksi)*
77(25)/Aluminium[b]	2·0
500(260)/Aluminium[b]	2·2
550(288)/Aluminium[b]	1·9
600(316)/Aluminium[b]	0·5
77(25)/V378A Graphite composite[c]	1·8
450(232)/V378A Graphite composite[c]	2·2
500(260)/V378A Graphite composite[c]	1·8
550(288)/V378A Graphite composite[c]	1·8
72(22)/PMR-15 Graphite composite[d]	1·6
400(204)/PMR-15 Graphite composite[d]	1·6
500(260)/PMR-15 Graphite composite[d]	1·5
600(316)/PMR-15 Graphite composite[d]	0·3
680(360)/PMR-15 Graphite composite[d]	0·16

[a] XEA9673 weighs 0·1 lbs/sq. ft and is supported on a glass scrim fabric. Cure cycle used was 1 h at 350°F (177°C)/25 psi (1·7 bar) pressure, followed by a post-cure of 2 h at 475°F (246°C) (unrestrained) for all substrates except PMR-15 graphite composite. The latter was post-cured for 2 h at 550°F (288°C).
[b] The aluminium adherends are FPL etched bare 2024-T3.
[c] V378A graphite composite is a US Polymeric bismaleimide resin prepreg on Union Carbide T300/3K unidirectional graphite.
[d] PMR-15 graphite composite is a NASA, Lewis polyimide resin prepreg on Celanese Celion woven graphite fibre.

addition to this product there exists a paste adhesive called XEA9369 which has fairly similar adhesive properties to XEA9673, but also is workable at ambient conditions unlike EA9351. The 'unique curing mechanism' technology is claimed to lead to resins with 600°F (316°C) capabilities, and with processing similar to conventional heat-cured epoxies.[74]

In the case of both of the Hysol product lines it should be noted that ease of processing is stressed. For example, for both EA9655 and XEA9673 the cure cycle consists of 1 h at 350°F (177°C) at 25 to 50 psi (3 to 6 bar) pressure, followed by an unrestrained cure of 2 h at 475°F (246°C) in an oven. This type of cure cycle is well within the bounds of presently used industrial equipment.

With the great interest in higher service temperature requirements in both electronics and aerospace, the matrix composites which are being developed will need to be fastened. Adhesive bonding provides an economical and reliable procedure. However, considerably more work and products are needed, especially to develop adhesives with epoxy-like processability and long term (1 to 10 year) thermal capabilities above 500°F (260°C). Bismaleimides may be able to provide such properties if properly formulated.

2.5.2. Electronic Uses for Bismaleimide Resins

Just as for adhesive technology the patent literature contains many references to use of modified bismaleimide resins for making laminates suitable for printed circuit boards. One example cited earlier in this review was the use of Kerimid 601 to make multilayer printed wiring boards.[43] Amongst the most frequently named companies engaged in research in this field are Hitachi Ltd, Rhone–Poulenc, Mitsubishi, Sumitomo Bakelite, Toshiba Chemical Products, Asahi Electrochemical, Tokyo Shibaura Electric, General Electric, Fujitsu Ltd, Mitsui Toatsu Chemicals, Inc. and Fujikura Cable Works Ltd.

The most cited reason for using bismaleimides refers to their heat-resistance to make good electrical insulators with dimensional stability. Some formulations are used as powder coatings, others as electrical coil insulants, encapsulants, moulding powders as well as adhesives for bonding copper to glass laminates. In many cases the bismaleimide resins are copolymers with epoxies or polybutadiene.

Normal processing in the electrical industry is different to that used in aerospace. The predominant method for making laminates is to cure them in a heated platen press. Cure temperatures are in the same range as for the previously described aerospace applications with pressures of 250–1500 psi (17–100 bar) for periods of 10 to 120 min. Such a cure may be followed by a post-cure of 3 to 5 h at 350–480°F (177–249°C).

The most commercially dominant resins that are supplied into the electrical/electronic market are based on Rhone–Poulenc's Kinel products. Mitsubishi BT resins are of great interest also.

In principle, any of the technologies discussed earlier are also applicable for electronic applications. Since there has been an explosion in the number of patents covering the use of bismaleimides in electronics (particularly in Japan), there is obviously much commercial interest in these materials.

REFERENCES

1. KAPLAN, S. L. and HIRSCH, S. S., *ACS Symposium Series, New Industrial Polymers,* **4,** 100 (1974).
2. HERTZ, J., *4th National SAMPE Conference,* Palo Alto, California, October, 1972.
3. BROWNING, C. E., The mechanisms of elevated temperature property losses in high performance structural epoxy resin matrix materials after exposures to high humidity environments, AFML-TR-76-153 (US Air Force Contract), 1976.
4. VADALA, E. T. and TRABOCCO, R. E., *13th National SAMPE Tech. Conf.*, p. 536, 1981.
5. BENSON DEXTER, H. and CHAPMAN, A. J., *12th National SAMPE Tech. Conf.*, p. 77, 1980.
6. ST CLAIR, A. K. and ST CLAIR, T. L., A review of high temperature adhesives, NASA Technical Memorandum 83141, July, 1981.
7. HERGENROTHER, P. M., *SAMPE Quarterly,* **3,** 1 (1971).
8. SERAFINI, T. T., Processable high temperature resistant polymer matrix materials, NASA TM X-71682, 1975.
9. SERAFINI, T. T., DELVIGS, P. and LIGHTSEY, G. R., *J. Appl. Polym. Sci.*, **16,** 905 (1972).
10. ST CLAIR, T. L. and JEWELL, R. A., *23rd National SAMPE Symp.*, p. 520, 1978.
11. LUBOWITZ, H. R., *ACS Org. Coatings and Plastics Chemistry,* **31** (1), 561 (1971).
12. LUBOWITZ, H. R., US Patent 3,528,950 (1970).
13. SHEPPARD, C. H., *29th National SAMPE Symp.*, p. 381, 1984.
14. KLOETZEL, M., *Organic Reactions,* Vol. IV (Roger Adams, ed.), Wiley, New York, p. 9, 1948.
15. ALDER, K. and TRIMBORN, W., *Ann. Chem.*, **566,** 58 (1950).
16. DYNES, P. J., PANOS, R. M. and HAMERMESH, C. L., *J. Appl. Polym. Sci.*, **25,** 1059 (1980).
17. LANDIS, A. L., BILOW, N., KELLER, L. B., BOSCHAN, R. H. and CASTILLO, A. A., *ACS Polymer Preprints,* **19** (2), 23 (1978).
18. TEDESCO, J. S. and SEIBOLD, R. W., Thermid IP-600. a new processible high-temperature polyimide resin, *Seminar on High-Temperature Performance Composites and Adhesives,* SAMPE, Inland Empire Chapter, June 16, 1983.
19. JONES, J. S., *21st National SAMPE Symp.*, p. 438, 1976.
20. SERAFINI, T. T., DELVIGS, P. and ALSTON, W. B., *27th National SAMPE Symp.*, p. 320, 1982.
21. STANDARD, C. M., VAN SICLEN, R. C. and PINKERTON, W. G., *28th National SAMPE Symp.*, p. 1095, 1983.
22. STOTLER, C. L., *25th National SAMPE Symp.*, p. 176, 1980.
23. BLECHERMAN, S. S., *25th National SAMPE Symp.*, p. 403, 1980.
24. FREEMAN, W. T. (Jr) and BAUCOM, R. M., *25th National SAMPE Symp.*, p. 418, 1980.
25. LOTTRIDGE, L., *25th National SAMPE Symp.*, p. 765, 1980.

26. SEARLE, N. E., US Patent 2,444,536 (1948).
27. ELLIOT, Y. R. (Ed.), *Macromolecular Synthesis,* Vol. 2, John Wiley Sons, Inc., New York, p. 111, 1966.
28. MANDGAL, S. and ST CLAIR, T. L., *SAMPE Quarterly,* **15** (3), 24 (1984).
29. COLE, N. D. and GRUBER, W. F., US Patent 3,127,414 (1964).
30. STENZENBERGER, H. D., *Applied Polymer Symp.,* **22,** 77 (1973).
31. STENZENBERGER, H. D., HEINEN, K. U. and HUMMEL, D. O., *J. Polymer Sci.,* **14,** 2911 (1976).
32. JONES, R. J., O'RELL, M. K. and SUTHERLAND, J. D., US Patent 4,116,937 (1978).
33. GRUNDSCHOBER, F. and SAMBETH, J., US Patent 3,380,964 (1968).
34. D'ALLELIO, G. F., US Patent 3,929,713 (1975).
35. JONES, R. J., GREEN, H. E. and QUINLIVAN, S. C., US Patent 4,179,551 (1979).
36. ZAHIR, S. A-C. and RENNER, A., US Patent 4,100,140 (1978).
37. BARGAIN, M., COMBET, A. and GROSJEAN, P., US Patent 3,658,764 (1972) and Br. Pat. 1,190,718 (1968).
38. BARGAIN, M., US Patent 3,883,486 (1975).
39. BACH, H. C., US Patent 3,637,534 (1972).
40. BACH, H. C. and HINDERER, H. E., *Polym. Preprints,* **11** (1), 334 (1970).
41. RENNER, A., FORGO, I., HOFMANN, W. and RAMSTEINER, K., *Helv. Chim. Acta,* **61** (4), Nr.144,1443 (1978).
42. KING, J. J., CHAUDHARI, M. and ZAHIR, S., *29th National SAMPE Symp.,* p. 392, 1984.
43. DARMORY, F. P., *19th National SAMPE Symp.,* p. 693, 1974.
44. MALLET, M. A. J. and DARMORY, F. P., *ACS Symposium Series, New Industrial Polymers,* **4,** 112 (1974).
45. ALVAREZ, R. T. and DARMORY, F. P., *20th National SAMPE Symp.,* p. 253, 1975.
46. STENZENBERGER, H. D., HERZOG, M., RÖMER, W., SCHEIBLICH, R. and REEVES, N. J., *Br. Polym. Journal,* **15,** 2 (1983).
47. STENZENBERGER, H. D., Br. Pat. 1,501,606 (1976).
48. STENZENBERGER, H. D., US Patent 4,211,861 (1980).
49. STENZENBERGER, H. D., HERZOG, M., RÖMER, W. and SCHEIBLICH, R., NASA Report CR-152324, 1979.
50. TECHNICAL BROCHURES for M751 and H795, Technochemie, GmbH, Verfahrenstechnik, 1981.
51. STENZENBERGER, H. D., HERZOG, M., RÖMER, W., SCHEIBLICH, R., REEVES, N. J. and PIERCE, S., *29th SAMPE Symp.,* p. 1043, 1984.
52. TECHNICAL BROCHURE, High heat resistant polyimide resin: BT resin, Mitsubishi Gas Chemical Co., Inc., 1984.
53. BROWNING, C. E., HUSMAN, G. E. and WHITNEY, J. M., Moisture effects in epoxy matrix composites, in *Composites Materials: Testing and Design (4th Conference),* ASTM STP 613, p. 481, 1977.
54. HEDRICK, I. G. and WHITESIDE, J. B., *AIAA Conference,* San Diego, March 24–25, 1977.
55. HERCULES TECHNICAL BROCHURE, Magnamite graphite prepreg tape AS4/3501-5A, Nos. 856 and 842-1.

56. JONES, R. J., GREEN, H. E., MARKLES, O. F. (Jr) and HOM, J. M., *25th National SAMPE Symp.*, p. 520, 1980.
57. VAUGHN, R. W., O'ROLL, M. K. and BUYNY, R. A., *23rd National SAMPE Symp.*, p. 528, 1978.
58. BROWNING, C. E., *23rd National SAMPE Symp.*, p. 541, 1978.
59. NASH, H. C., PORANSKI, C. F. (Jr) and TING, R. Y., Effect of room-temperature ageing on graphite/polyimide prepreg materials, in *Resins for Aerospace* (C. A. May, Ed.), ACS Symp. Series 132, Article 34,469, 1980.
60. POVEROMO, L. M., *27th National SAMPE Symp.*, p. 729, 1982.
61. NASA CONTRACT 165339, Hybridized polymer matrix composites, Grumman Aerospace Corp. (1978–1980).
62. MCKAGUE, L., V378A polyimide resin—A new composite matrix for the 1980's in *Composites for Extreme Environments* (N. R. Adsit, Ed.), ASTM STP 768, American Society for Testing and Materials, p. 20, 1982.
63. STREET, S. W. and BECKLEY, D. A., US Patent 4,351,932 (1982).
64. STREET, S. W., V-378A; an addition polyimide for advanced composites, *First International Conference on Polyimides,* Society of Plastics Industry Conference, Ellenville, New York, Nov. 10–11, 1982.
65. STREET, S. W., *25th National SAMPE Symp.*, p. 366, 1980.
66. WILKINS, D. J., Naval Air Systems Command Report NAV-GD 0037, 1981.
67. MCKAGUE, L., *28th National SAMPE Symp.*, p. 640, 1983.
68. HOPPER, L. C. and HARRISON, E. S., *27th National SAMPE Symp.*, p. 23, 1982.
69. DATA BULLETIN R5254C-R1, Narmco Materials Inc., April, 1983.
70. HO, V., *29th National SAMPE Symp.*, p. 409, 1984.
71. HSU, M. S., PARKER, J. A., CHEN, T. S. and HEIMBUCH, A. H. *29th National SAMPE Symp.*, p. 1034, 1984.
72. TECHNICAL BULLETIN, BT resin adhesives for polyimide film, trial product A110, A120, Mitsubishi Gas Chemical Co. Inc., 1984.
73. LANDMAN, D., New processable polyimide based adhesives, *First Technical Conference on Polyimides,* SPE. Inc., Ellenville, New York, Nov. 10–11, 1982.
74. LANDMAN, D., *28th National SAMPE Symp.*, p. 740, 1983.

Chapter 3

REVIEW OF PREPREG TECHNOLOGY

GOTTFRIED WIEDEMANN and HORST ROTHE

Academy of Sciences of the German Democratic Republic, Institute for Technology of Polymers, Dresden, German Democratic Republic

SUMMARY

This chapter discusses the prepreg process in fibre–resin composites. The central feature of the process is the production of a partially crosslinked state in the resin matrix, and this is discussed in the context of the total polymer processing operation. The concept of a prepreg is a broad one, but this chapter focuses on 'classical' materials, and the problems described relate to electrical grade laminates, which have to meet extremely strict requirements. The raw materials (reinforcement and resin) are briefly outlined, together with an account of the behaviour of prepregs during subsequent manufacturing, especially at the pressing stage.

The principal emphasis is on prepreg formation, and on the characterisation of properties, together with various technological problems. These latter relate mainly to impregnation and to the supply of heat within the dryer. Attention is given to undesirable defects occurring in prepregs, and to technological calculations relating to weight, and to pre-curing. Computer-aided control is also mentioned.

The extent of pre-cure is important, and several diagnostic test methods are discussed, along with the effect of the state of pre-cure on behaviour in the press, as well as on the final material properties. Trials to simulate the many influences on prepreg properties are briefly outlined. Finally, probable future trends are discussed. A dramatic change in prepreg processing technology is not expected, but innovations will continue to produce improvements.

3.1. INTRODUCTION

The term 'prepreg' was derived from the more cumbersome expression 'pre-impregnated material'. When the technology was first developed in the early fifties the term generally concerned materials based on woven fabric and epoxy resin. Nowadays, the range of available reinforced materials is much wider, and the term has been used for pre-impregnated paper,[1] for reinforcements impregnated with thermoplastics,[2] and for sheet moulding materials (SMC).[3] All these materials are produced with special technological equipment.

However, these technologies and materials are so diverse that we shall simply concentrate on the 'classical' prepreg process, i.e. applying a solution by dipping the reinforcement to impregnate it, and then using a vertical dryer to evaporate the solvents and pre-cure, or partially cure, the resin. Figure 3.1 is a schematic illustration of the process.

The term 'prepreg' will therefore be taken to mean an intermediate product, consisting of reinforcement and partly crosslinked polymer, i.e. pre-cured resin, suitable for shaping to produce a fabricated

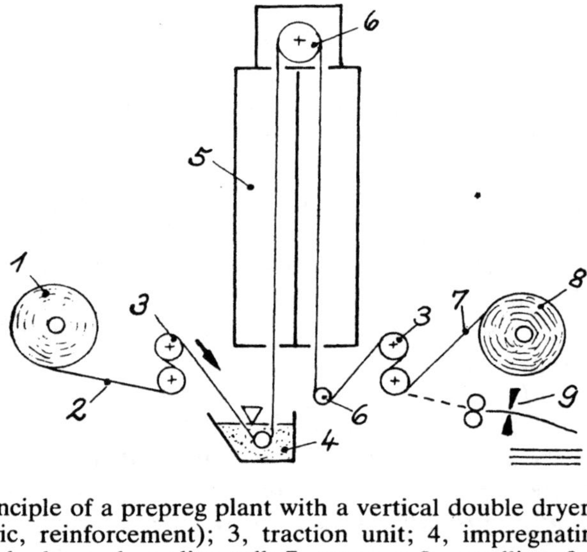

FIG. 3.1. Principle of a prepreg plant with a vertical double dryer. 1, Unwind; 2, web (fabric, reinforcement); 3, traction unit; 4, impregnating tank (wet end); 5, double dryer; 6, cooling roll; 7, prepreg; 8, uprolling; 9, cross cutter.

reinforced plastics product. The term 'intermediate product' applies to the compound at the stage when it is ready for shaping. The material is at an intermediate stage of its curing reaction, when hardener and accelerator have been added to the polymer, and a certain amount of curing has occurred.

The commonest reinforcement is continuous 'E' glass fibre. Recently, aramid and carbon fibres have progressively come into use, not only in homogeneous or one-fibre systems but also in hybrid reinforcements.

With regard to forms, fabrics (especially woven fabrics) have dominated up to now, but other textile forms are increasingly attractive, such as uni-, bi- or tri-directional fabrics; stitch-bonded threadlayer fabrics, and non-wovens. Linear reinforcements are available for special applications, e.g. threads, rovings and tapes.

Epoxy and polyester resins are the preferred matrix materials, but use is also made of phenolics, melamines, silicones, polyimides, and polyurethanes, among others. In addition to fibres and resin, the prepreg contains hardener, accelerator and remains of solvent. There may also sometimes be thermoplastics, elastomers and fillers present.

The shaping of reinforced plastics (RP) products using prepreg is preferably done by laminating, using heated multi-daylight presses. Moulding tools are applied to produce formed pieces. Winding processes are also available for prepregs.

The properties of prepregs have to be adapted to these processing operations, and the sections following will deal with these matters.

The use of prepregs has many advantages. The resin-to-fibre ratio can be controlled, and the prepregs are ready to use, with no further chemicals, to allow simple shaping operations. Problems of automation can also be more readily solved. On the other hand, manufacturing with prepregs requires closer co-operation with the prepreg producer. The co-operation must relate to choice of reinforcement, choice of resin type, and the organisation of the production schedule, because the shelf-life of prepregs is limited; crosslinking continues to occur during storage, at a rate which depends very much on temperature. The general principles of using reinforced plastics apply also to prepregs. Mechanical strength depends on the reinforcement—its type, content and arrangement. The chemical, electrical, and thermal properties result from the selection and correct formulation of the resins with their various additives.

Finally, some important application areas for prepregs should be mentioned. They are chiefly:

(a) laminates for electrical insulation,
(b) base materials for printed circuit boards (PCBs), and
(c) load-bearing components for various purposes such as sports goods, vehicles, aircraft, chemical process equipment, etc.

Base materials for PCBs have to meet extremely demanding specifications. For this reason the problems discussed later in this chapter are often taken from this field of application.

3.2. PREPREG FORMING—AN INTEGRAL PART OF THE TOTAL POLYMER PROCESSING OPERATION

Prepreg forming is an integral element of polymer processing. It is hoped to achieve the following targets.

(a) Concerning the *material shaping:*
—composite-forming without any defects,
—realisation of a pre-determined prepreg weight, so as to achieve a certain laminate quality.
(b) Concerning the *chemical conversion:*
—limited volatile content,
—realisation of a given state of crosslinking to achieve a certain flow during press-processing.

3.2.1. Composite-forming without Defects

The composite is formed during the impregnation process, which influences the prepreg quality. A perfect prepreg should have even surfaces, the polymer matrix should not contain any voids or impurities, and the contact between polymer and reinforcement should be a complete one. This is desirable to achieve dimensional stability[4] and to prevent penetration of liquid media into the interior of the laminate.[5,6]

Some defects and other phenomena, occurring in prepregs are discussed below, together with their causes and effects on the laminates.

One important kind of surface defect, characterised by exposed reinforcement, occurs if the resin content is too low. It can occur either locally or generally, and is called ‘weave texture’ or ‘weave exposure’.[7] Another problem is that prepreg can contain many

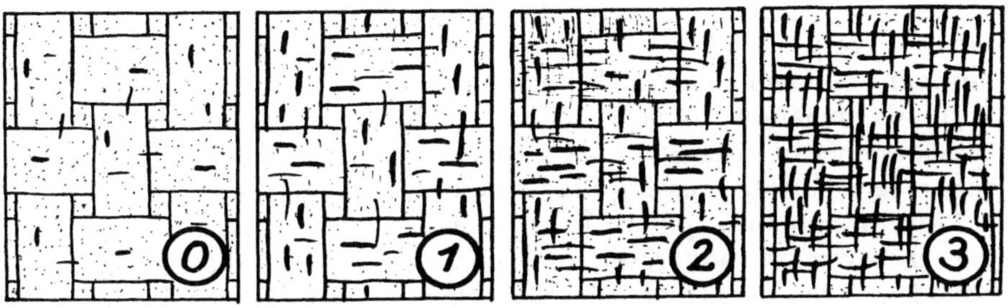

FIG. 3.2. Sketches of prepregs containing 'capillary lines' of graduated intensity, valued by marks as follows: Mark 0, 'not and hardly', resp., to be seen; mark 1, 'slightly' to be seen; mark 2, 'much' to be seen; mark 3, 'very much' to be seen.

pinholes if drying is carried out too quickly. If the *B*-time is too short then the pinholes cannot be completely eliminated by subsequent pressing and the result is a voided laminate.

Often prepregs contain so-called 'capillary lines'. Figure 3.2 illustrates the phenomenon schematically. There are different degrees of seriousness, indicated by marks 0, 1, 2 and 3. The capillary lines can be observed at low magnification and they run within the filament

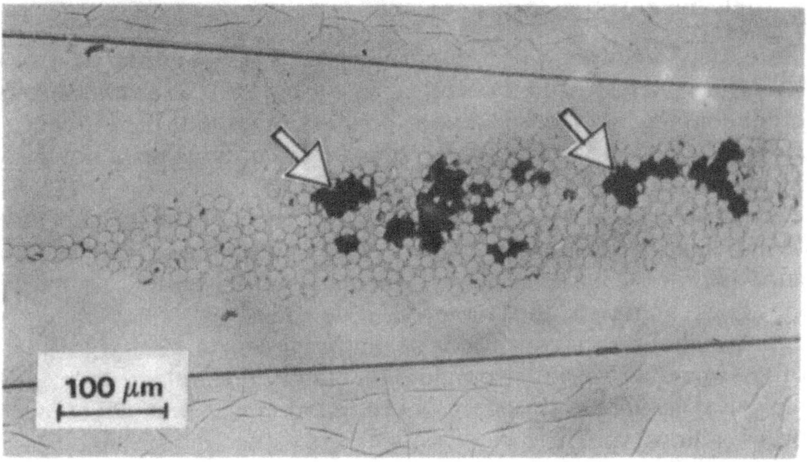

FIG. 3.3. Polished cross-section of a prepreg containing capillary lines. Within the filament bundles there are voids (microcavities, channel-like) not filled up with polymer, see arrow. On polishing these voids have grown larger, note the cracked circular areas.

yarns of warp and weft. Their diameter is about 5 to 50 μm and they are not very long. We ascertained that capillary lines are voids between the filaments, which have not been filled with resin (see Fig. 3.3). Examination of a laminate made from prepreg containing capillary lines of intensity mark 1 to 2 shows that there were no hollow spaces left. During pressing they had filled up with resin.

More prominent capillary lines (mark 3) do not vanish on laminating; they remain, and result in a cross-section like that in Fig. 3.3. Such laminates have to be rejected as high quality materials because of the possibility of fluid penetration and perhaps short circuits may happen if the laminate is applied to electrical use. Such an effect would be very undesirable.

The better the quality of the impregnation, the better the laminate. So what does the quality of the impregnation depend on? Using Hagen–Poiseuille's law, and applying it to the impregnation process[8] by considering the process as essentially the filling of capillary systems by liquids, we see that the quality of impregnation depends:

(a) *directly* on the surface tension of the resin mixture, the cosine of the contact angle, and the time of contact (or alternatively, the reciprocal of the web speed), and
(b) *indirectly* on the viscosity of the polymer mixture and on the thickness of the reinforcement.

These dependences were confirmed by investigations based on epoxy resin systems, using several types of woven fabrics finished with different coupling agents.[8] It should be mentioned that the influence of viscosity is not so striking with the range of viscosities used nowadays, i.e. up to about 50 mPa s.

The fabric weight used for prepregs ranges from about 50 to 250 g/m^2, and therefore the thickness of a ply varies widely (from about 0·05 to 0·3 mm). Usually, increasing the fabric weight also implies thicker threads and higher yarn densities.

Light weight fabrics can easily be impregnated to give high quality prepregs (mark 0 to 1). Applying the same technology, i.e. single dipping, to heavier fabrics is more economical but gives worse prepregs (mark 2 or 3).

The possibility of improvement in impregnation quality by other means is discussed further in Section 3.4.1.

All the factors mentioned above influence the impregnation but capillary lines cannot simply be attributed to a 'bad wetting' of the

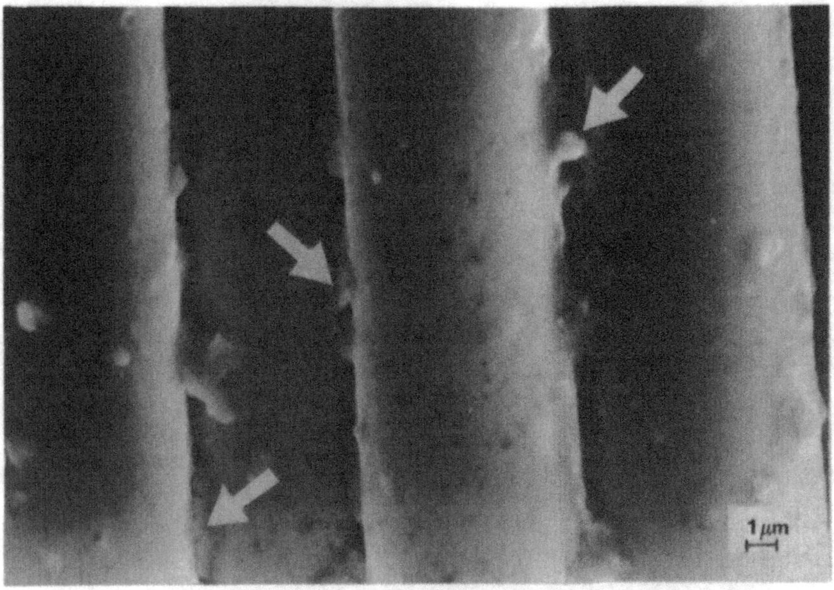

FIG. 3.4. Detail of a finished glass fabric—filaments are gummed up by coupling agent, see arrows.

reinforcement with the resin mixture. There are two factors which hinder the penetration of resin mixture into the fabric:

(a) The thicker the fabric, the bigger the air volume within the compactly arranged filaments, which must be displaced by the resin mixture.
(b) Filaments are gummed up by coupling agent (see Fig. 3.4) resulting in microcavities.

3.2.2. Calculation of the Prepreg Weight

By experience it is known that prepregs with too low a resin content result in laminates with uneven surfaces. This is called 'weave texture' and 'weave exposure'. For high quality requirements, e.g. base materials for printed circuit boards, laminates like this are unacceptable. On the other hand, if the resin content is too high, there is a risk of the 'swimming out' of prepreg sheets during the press-processing and such a laminate is absolutely useless.

Therefore it is necessary to produce prepregs within a certain resin content, or with a certain prepreg weight for a given fabric. Trials for

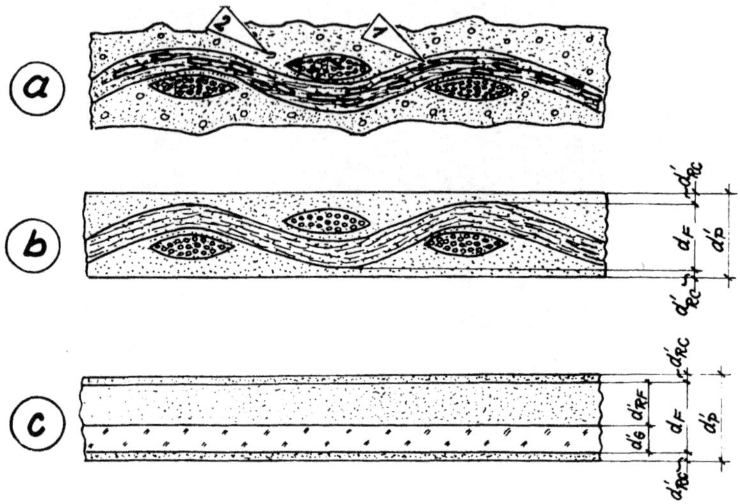

FIG. 3.5. Cross-section through a prepreg. (a) Really—containing cavities (1) and bubble-like voids (2), surfaces are uneven; (b) ideally—without any cavities and voids, surfaces are smooth; (c) schematically—consisting of layers for glass (G), resin within the fabric (RF), and resin coatings at both sides (RC).

calculating the prepreg weight, based on a constant resin content, showed results which could not be confirmed by experience. This was caused by the fact that the thickness of a fabric depends not only on its weight but also on its textile structure.[9] In this case, the thickness has to relate to the unit pressure applied when pressing. Besides, there are matrix resin effects, e.g. matrix density, which have to be taken into consideration as well.

Figure 3.5(a) shows a cross section through a prepreg: the surfaces are uneven, and voids or microcavities within polymer and fabric exist.

Figure 3.5(b) illustrates the same prepreg in idealised form; the surfaces are smooth, and all voids and cavities are filled with polymer. A 'compressed prepreg' like this can be compared with a 'one-layer-laminate', pressed without any flow.

The 'effective thickness' of a prepreg d'_P is given by

$$d'_P = d_F + 2d'_{RC} \qquad (1)$$

where d_F is the fabric thickness when pressure is applied, and d'_{RC} is the effective thickness of the resin coating on both sides of the prepreg.

d_F is composed of both the effective thicknesses: d'_G for glass, and d'_{RF} for the resin within the fabric filling up all the pores. This is schematically illustrated in Fig. 3.5(c).

Then d'_P is given by

$$d'_P = d'_G + d'_{RF} + 2d'_{RC} \tag{2}$$

The three coats of resin can be treated as having an effective total resin thickness d'_R and therefore d'_P is given by

$$d'_P = d'_G + d'_R \tag{3}$$

The thickness d of a sheet is generally given by

$$d = m/\rho \tag{4}$$

where

m = weight per unit area

ρ = density of the material

As the weight of the resin m_R is the difference between weight of prepreg m_P and weight of glass fabric m_G

$$m_R = m_P - m_G \tag{5}$$

the effective thickness of the prepreg d'_P can be calculated

$$d'_P = \frac{m_G}{\rho_G} + \frac{m_P - m_G}{\rho_R} \tag{6}$$

where ρ_G is the density of glass and ρ_R the density of resin.

Analogous to eqn (2) the prepreg weight m_P can be expressed

$$\underset{\text{weight of prepreg}}{m_P} = \underset{\text{fabric}}{m_G} + \underset{\text{resin within the fabric}}{\rho_R\left(d_F - \frac{m_G}{\rho_G}\right)} + \underset{\text{resin coatings}}{2\rho_R d'_{RC}} \tag{7}$$

Based on eqn (6) the laminate thickness d_L (free of any voids) is given by

$$d_L = n\left[\frac{m_G}{\rho_G} + \frac{m_P - m_G}{\rho_R}(1 - \phi_R)\right] \tag{8}$$

where n is the number of prepreg sheets and ϕ_R the resin scrap at the press stage.

3.2.3. The Partly Crosslinked State (State of Pre-cure)

3.2.3.1. B-Stage

Prior to cure solvents have to be removed by evaporation to give a volatile content of less than 0·5%.

The curing itself is a chemical reaction which, for a thermosetting polymer, such as an epoxy resin, can (according to Enns and Gillham[10,11]) be understood in terms of a time–temperature transformation (TTT) as shown in Fig. 3.6. Both the times for gelation and vitrification are plotted against the isothermal cure temperature. There are four distinct regions: liquid, gelled rubber, ungelled glass, and gelled glass. The boundaries are defined by the S-shaped vitrification curve and the gelation curve. The glass transition temperatures marked in this diagram are called: T_g (resin) for the unreacted resin mixture; T_g (gel) for the resin at its gelation point; and $T_g\ \infty$ for the fully cured resin. Gelation results from the crosslinking reaction, which brings about conversion from the liquid to a solid state. It can be attributed to the formation of branched molecules, and is associated with a very rapid increase in viscosity, as discussed further in Chapter 5. The crosslinks consist of more or less flexible polymeric segments and as the chemical reaction produces more such segments the physical nature of the material changes such that the material becomes glassy.

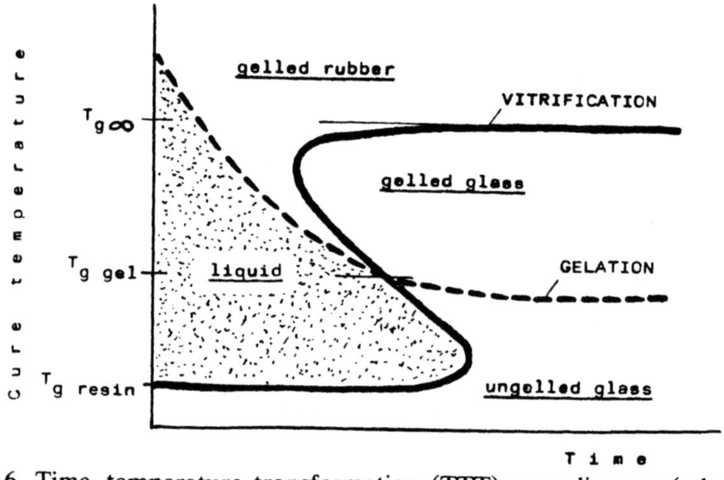

FIG. 3.6. Time–temperature transformation (TTT) cure diagram (schematic) according to Enns and Gillham.[11]

This chemical conversion from liquid via (gelled) rubber up to the (gelled) glass state is called 'pre-curing'. In the case of the prepreg, a state between gelation and vitrification is 'frozen' by cooling down. This partly crosslinked state, called the 'B-stage', is defined to be a range but not a point.

Furthermore, a certain B-stage, expressed as 'B-time of the prepreg' is intended to achieve a desirable 'flow-curing behaviour' during pressing.[9] The difference between the B-time of the resin mixture and that of the prepreg is called the 'B-time reduction', and will be calculated in the next section. The longer the B-time reduction proceeds, the more the crosslinking goes ahead.

3.2.3.2. Calculation of B-time Reduction

Later, the thermal parameters of the prepreg process will be related to the reactivity of the resin mixture based on Arrhenius' law. The activation energy E is given by

$$E = R\frac{T_1 T_2}{T_2 - T_1} \ln \frac{k_1}{k_2} \tag{9}$$

where R = gas constant, T_1, T_2 = temperatures of reaction, and k_1, k_2 = reaction rates.

For the given problem the reaction rates can be substituted by B-times B_{t_1} and B_{t_2}, and with $\ln x = M \log x$ ($M = \ln 10 \approx 2{\cdot}30$) E can be written as

$$E = RM\frac{1}{\dfrac{1}{T_1} - \dfrac{1}{T_2}} \log \frac{B_{t_1}}{B_{t_2}} \tag{10}$$

and furthermore

$$E = RMK \tag{10a}$$

where K is a temperature gradient

$$K = \frac{\log B_{t_1} - \log B_{t_2}}{\dfrac{1}{T_1} - \dfrac{1}{T_2}} \tag{11}$$

called a 'K-factor', which is a characteristic value proportional to the activation energy.

In practice, the K-factor is determined by measuring the B-time at

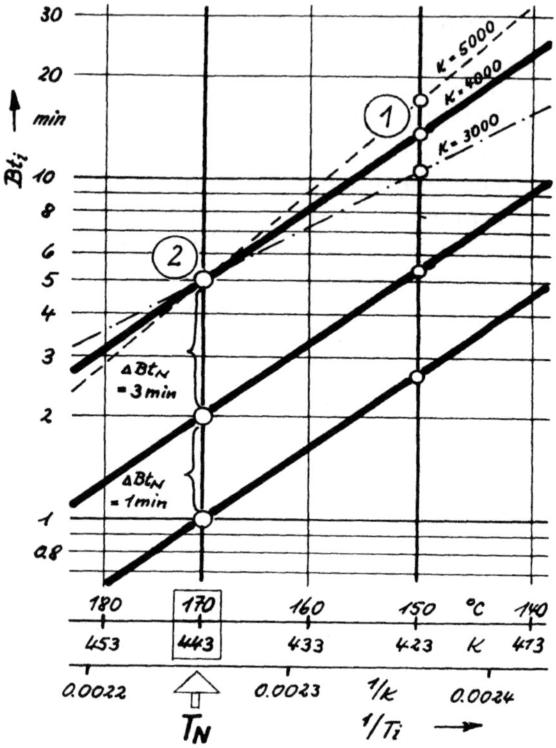

FIG. 3.7. Straight K-factor lines—logarithm of B-time equals the reciprocal of absolute temperature.

two temperatures, and plotting the value couples in a half-logarithmic scale (see Fig. 3.7). With regard to this diagram B-time reduction is equal to a parallel traversing of the 'K-factor straight line'.

As the B-time depends on temperature, all measurements have to relate to a so-called 'normalised temperature' T_N (e.g. $T_N = 443\,K \triangleq 170°C$). Subsequently, two B_t subscripts are used: N indicates T_N, and i relates to any temperature T_i.

The conversion from B_{t_N} to B_{t_i} is given by

$$\log B_{t_i} = K\left(\frac{1}{T_i} - \frac{1}{T_N}\right) + \log B_{t_N} \tag{12}$$

Based on this equation, the differential of B-time reduction dB_{t_N} can be calculated for exposing a resin mixture or prepreg, respectively, at

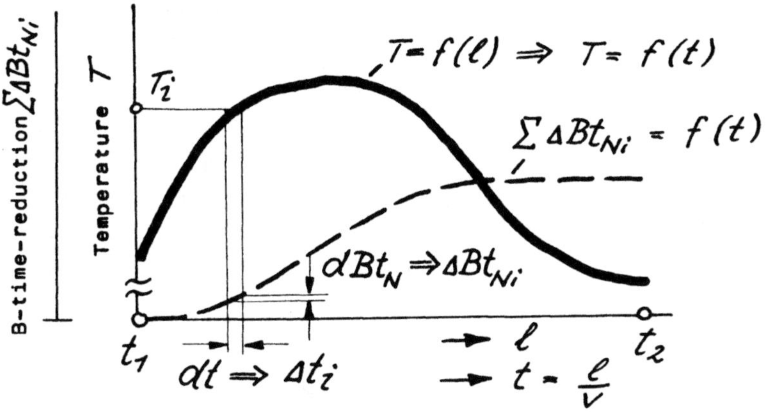

FIG. 3.8. Schematic temperature profile, and effected B-time reduction when a prepreg is passing.

temperature T_i during a time interval dt

$$dB_{t_N} = 10^{-K[(1/T_i)-(1/T_N)]}\,dt \tag{13}$$

That is the base to calculate the B-time reduction which is a target for the prepreg process, depending on the K-factor, and on conditions of thermal history.

When passing the dryer, the web is exposed to a temperature–length profile $T=f(l)$ which can be expressed as a temperature–time profile $T=f(t)$ when referring to a certain web speed v (see Fig. 3.8).

In principle, the B-time reduction ΔB_{t_N} can be calculated between the time limits t_1 and t_2 by integration

$$\Delta B_{t_N} = \int_{t_1}^{t_2} 10^{-K\{[1/T=f(t)]-(1/T_N)\}}\,dt \tag{14}$$

As the analytical expression for $T=f(t)$ is unknown, for practical applications this calculation is done 'step-by-step'

$$\Delta B_{t_N} = \sum \Delta B_{t_{N_i}} = \sum_{i=1}^{j} \frac{\Delta l}{v} 10^{-K[(1/T_i)-(1/T_N)]} \tag{15}$$

where Δl is the distance between the thermocouples arranged along the web run, v is the web speed, and T_i is the temperature of each step.

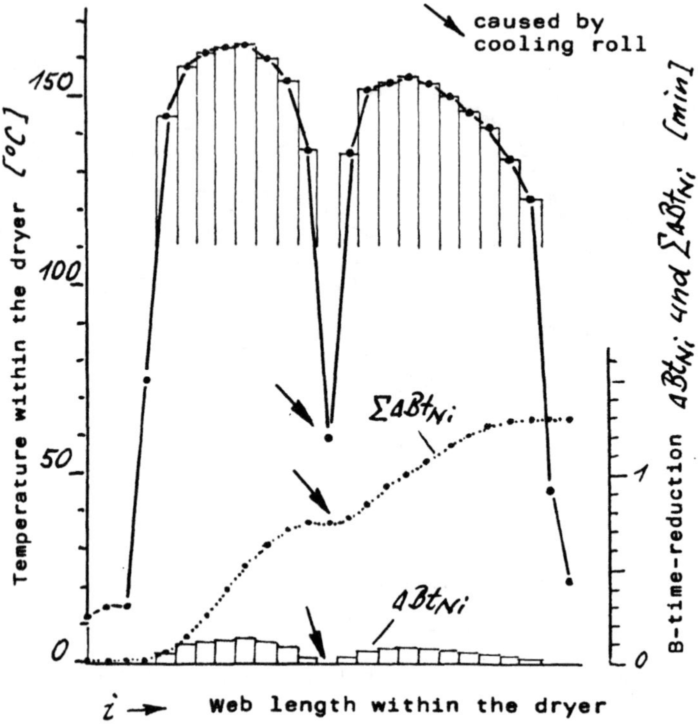

FIG. 3.9. Fairly typical temperature profile of a vertical double-dryer and effected B-time reduction.

Applying eqn (15) to a temperature profile of a vertical dryer (see Fig. 3.9 as an example) the B-time reduction was calculated for each step ($\Delta B_{t_{N_i}}$) and plotted, and the sum-curve ($\sum \Delta B_{t_{N_i}}$) as well. The strictly increasing influence of temperature to B-time reduction is apparent. Using this formula, it is possible to estimate the influence of technological parameters, e.g. temperature profile, web speed, resin reactivity (K-factor), on the B-time reduction.

At present, usually, the temperature of a dryer is measured at only one or a few points to control the heat supply, i.e. to influence the temperature profile. This procedure often results in variations of B-time reduction, causing variations in the prepreg B-time. Hence, the material can behave differently during pressing, especially with regard to flow-curing conditions.

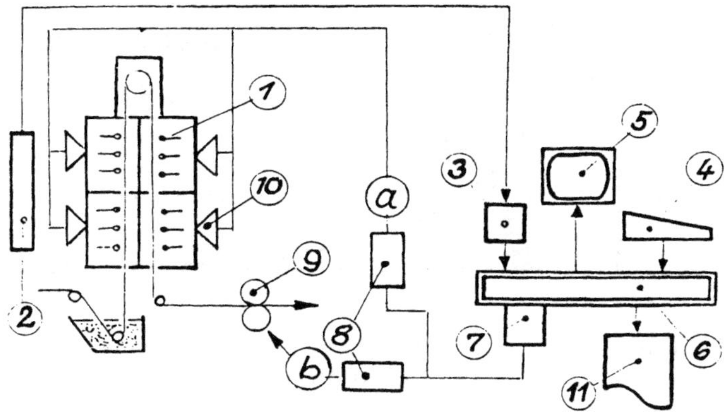

Fig. 3.10. Scheme of prepreg process with a computer-aided control plant. 1, Thermocouple; 2, central recorder for temperatures; 3, conversion analogous/ digital; 4, keyboard; 5, picture screen; 6, microprocessor (central unit); 7, controller; 8, control element; 9, traction unit; 10, heating unit; 11, listing.

In modern prepreg plants, the temperature profile is continuously measured to achieve a computer-aided control. A control plant of this kind is illustrated schematically in Fig. 3.10. The B-time reduction, based on the reactivity of the resin mixture and the temperature profile, is continuously calculated. Comparing this result with the intended value, the heat supply (a) and web speed (b), respectively, are controlled. This is done to increase the efficiency and yield by operating at higher temperatures and web speeds and nevertheless, the extent of crosslinking can be controlled.

3.2.3.3. Calculation of Curing Rate

The calculation of B-time reduction, as mentioned above, is good for technological use but it does not describe the curing of the resin sufficiently. As discussed below, the curing process can be predicted from the DSC parameters of the resin mixture, and relations to B-time reduction can be deduced.

The curing of epoxy resins can be described as a reaction of the first order; that means an isothermal cure, expressed by the so-called 'curing rate', proceeds like an exponential function (see Fig. 3.11).

The curing rate, α, given by

$$\alpha = 1 - \exp[-k_i t] \tag{16}$$

tends to unity as time t tends to infinity, whereas the first derivative

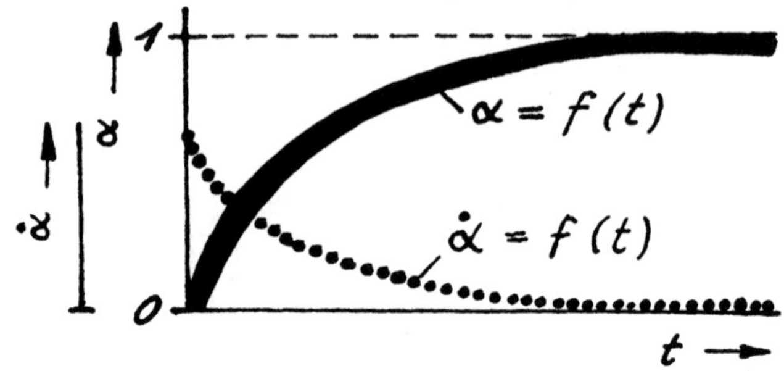

FIG. 3.11. Curing rate and its first derivative equal the time at isothermal exposure.

$d\alpha/dt = \dot{\alpha}$ tends to zero

$$\dot{\alpha} = k_i \exp[-k_i t] \tag{17}$$

where k_i, the reaction rate at any temperature of exposure T_i, is calculated, according to Arrhenius, by

$$k_i = k_N \exp\left[-\frac{E}{R}\left(\frac{1}{T_i} - \frac{1}{T_N}\right)\right] \tag{18}$$

where
E = activation energy
R = gas constant
T_N = reference temperature (e.g. 443 K ≙ 170°C)
k_N = reference reaction rate

Analogous to the calculation of B-time reduction, by eqn (15), the curing rate of the prepreg α_P can be deduced

$$\alpha_P = 1 - \exp\left[-\Delta t \sum_{i=1}^{j} k_i\right] \tag{19}$$

and combined with $\Delta t = \Delta l/v$ and eqn (18)

$$\alpha_P = 1 - \exp\left[-\frac{\Delta l}{v}\sum_{i=1}^{j} k_N \exp\left[-\frac{E}{R}\left(\frac{1}{T_i} - \frac{1}{T_N}\right)\right]\right] \tag{20}$$

Using this double exponential function the curing rate can cumulatively be calculated proceeding step-by-step from thermocouple 1 to j.

3.2.3.4. Relation between B-time Reduction and Curing Rate

To find a relation between the B-time reduction ΔB_{t_N} and the curing rate of the prepreg, α_P, two special states of partial crosslinking (pre-curing) are to be considered, see Fig. 3.12.

(a) The curing rate, α_B, which exists when the resin stops being 'ropy' at the B-time test, marked with (a).
(b) The curing rate of prepreg, α_P, marked with (b).

Based on eqn (17) an integral equation can be written

$$\alpha = \int k_i \exp[-k_i t]\, dt \tag{21}$$

and using limits t_1 and t_2 this formula results in

$$\alpha = \exp[-k_i t_1] - \exp[-k_i t_2] \tag{22}$$

To the B-time of mixture $B_{t_{NM}}$ the limits $t_1 = 0$ and $t_2 = t_B$ are valid:

$$B_{t_{NM}} = -\frac{1}{k_N} \ln(1 - \alpha_B) \tag{23}$$

And with the limits $t_1 = t_P$ and $t_2 = t_B$ the B-time of prepreg as function of both the curing rates can be expressed

$$B_{t_{NP}} = -\frac{1}{k_N} \ln \frac{1 - \alpha_B}{1 - \alpha_P} \tag{24}$$

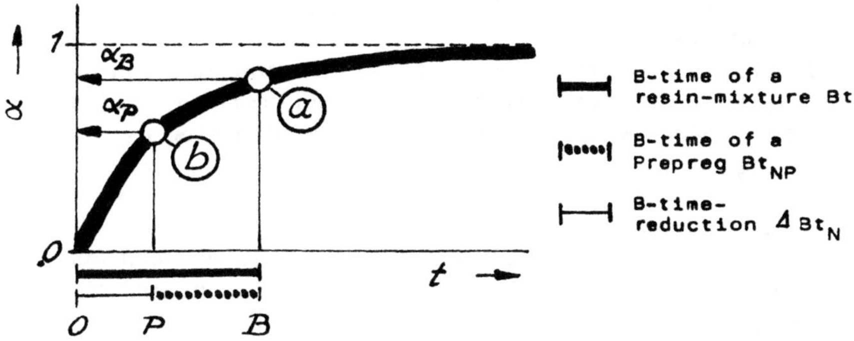

FIG. 3.12. Curing rate α versus time t. (a) State of curing at the end of B-time testing; (b) state of pre-curing typical to a prepreg.

For technological considerations the relative B-time reduction z can be useful (see Section 3.3.3.2)

$$z = 1 - \frac{B_{t_{NP}}}{B_{t_{NM}}} \tag{25}$$

and expressed in terms of curing rates

$$z = \frac{\ln(1-\alpha_P)}{\ln(1-\alpha_B)} \tag{26}$$

where the meaning of the subscripts is: P = prepreg, M = mixture, B = state of curing at B-time test, N = 'normalised' temperature.

Finally, combining eqns (15), (20) and (10a) we get the formula

$$\alpha_P = 1 - \exp\left[-k_N\, \Delta B_{t_N}\right] \tag{27}$$

where k_N is the reference reaction rate, and ΔB_{T_N} the B-time reduction. Notice that in this equation, the curing rate of the prepreg, α_P, is independent of activation energy and temperature.

3.3. CHARACTERISATION OF PROPERTIES

3.3.1. Raw Materials

The characterisation of prepreg properties is important in their manufacturing and for developments in this field as well. It begins with the testing of the raw materials: i.e. the reinforcements (fabrics) and the polymer (resin).

With fabrics the testing relates to characteristic data and tolerances such as weight, thickness, width, textile construction (yarns, density, weave), and finish.

Very important for prepreg properties is the testing of the properties of the resin and the resin mixture, which should concern, at least: the epoxide equivalent, solid content, viscosity, solubility, purity, state of crosslinking (B-time), and reactivity (K-factor).

3.3.2. Test Methods

For several years scientific methods have increasingly been applied, e.g., DSC (differential scanning calorimetry), TMA (thermal mechanical analysis), DMA (dynamic mechanical analysis), IR (infrared spectroscopy), FT–IR (Fourier-transform IR-spectroscopy), GPC (gel permeation chromatography), and, last but not least, HPLC (high pressure liquid chromatography).[12] These methods are valuable not only for checking the identity of deliveries or charges, but also to give characteristics for calculation of *B*-time reduction and curing rate, and to predict or influence the processing (onset, temperature, and duration of curing).

These modern methods are also increasingly applied for improving the prepreg process, and they are suitable instruments for a scientific analysis of the total process, too.[13] But at present they are not yet developed to perfection as it is necessary to reflect the totality of physical, chemical, and technological procedures in the prepreg process.

Several years ago Persson,[14] a specialist in prepregs and laminates, said that carrying out these processes is 'not only a science but also an art'. Today, this statement is still partly true. Thus, a certain 'empirical approach' has its place too. Besides the scientific methods, as mentioned above, several tests which appear simple and quick are still applied for checking prepreg processes, e.g. *B*-time, viscosity using the Ford-cup,[15] softening point according to Kofler,[16] etc.

There is a further aspect to employing modern scientific methods in industrial practice. These instruments are very expensive, and they require highly educated teams to operate them. These methods have not been available to all producers of prepregs until recently, for economic reasons.

3.3.3. Prepreg Properties

The prepreg properties are to be adapted to requirements both of the (existing) plants for manufacturing, e.g. press equipment, and of the specific properties to final products, e.g. resistance to soldering processes required for PCB's.[17–20]

These properties have to be measured by various different methods. Listed below are the most important characteristics—excluding those relating to the characterisation of pre-curing (see next section). The references also include details of the test methods being used:

(a) *Weight:* tested by weighing samples with a fixed area.[21–23]
(b) *Thickness:* tested at a fixed unit pressure.[21–23]
(c) *Resin content:* tested by combustion and weighing.[21–23]
(d) *Volatile content:* tested by drying and weighing.[21–23]
(e) *Stiffness:* the sample is fixed horizontally like an unrestrained beam and it is therefore bent by its own weight.
(f) *Dielectrical properties.*[24–26]
(g) *Flame resistance:* (for fire-retardant types only).[21,27]

Users of prepregs also have to take into consideration the limited *shelf-life* of prepregs, caused by the fact that the cross-linking proceeds continuously at room temperature, resulting in a decreased flow. Besides, they have to pay attention to the *catalytic effects* of radiation, ultra violet included.

Prepregs are '*climate sensitive*'[28] because they quickly absorb moisture, resulting in a decrease in T_g[6] which results in a poor resistance of laminates to fluctuating temperatures.[29] Therefore, for products of high quality, moisture content has to be considered, that means, condensation of water from humid air onto prepregs should be avoided, and drying might be necessary.

The prepreg state is characterised by the partially cross-linked state in the matrix, i.e. the polymer is only partly cured. Therefore the term 'pre-curing' shall be preferred. The following sections will deal with the characterisation of this state, which is the main influence on the pressing and on the properties of the final products. Both effects are to be discussed, and, last but not least, trials for a scientific analysis of the total process by mathematical simulations are presented.

3.3.3.1. Characterisation of the Pre-cure State

On examination of a series of prepregs of gradually increasing degree of cure, it is possible to detect a change in their feel and in their stiffness. At first they feel tough and sticky, and are flexible. Such materials are said to be 'soft'. At an advanced stage, however, they are brittle and stiff; such a state is summarised as 'hard'. Thus it is possible for experienced workers to gauge the state of pre-cure, roughly at least, by feel alone. But the characterisation of the state of cure is more precisely carried out by studying what happens when more heat is applied to continue the cure process, as follows:

(a) The *B-time,* or the gel-time, can be measured by using resin particles at a standard or 'normalised' temperature, and observing the ropy behaviour, or time to reach the state of ropiness.[30,31]

(b) *Initial softening temperature* (in °C). According to Kofler and to Boëtius this can be measured at a series of increasing temperatures by observing changes in the optical characteristics of the prepreg surface.[16,32]

(c) *Deformability*. This is measured by observing the changes in thickness as a function of temperature at constant load.[33]

(d) *Remaining reaction heat* (in J/g). This is determined by differential scanning calorimetry (DSC).[13,34]
(e) *Glass transition temperature* (in °C). This can be measured either by DSC[35] or by dynamic mechanical analysis, noting the damping characteristics.
(f) *Flow characteristics.* These are measured by a quick pressing test, i.e. placing biased samples of known weight in the press at a standard temperature and pressure, and observing the flow as the amount of scrap for a given weight.[21]

These methods can be applied to prepregs at the 'soft' and the 'hard' end of the pre-cure spectrum but the results can be confusing. Table 3.1 gives a survey relating to resin, prepreg and laminate. The latter will be discussed in the next section but one.

Besides the properties mentioned above further parameters should be mentioned since they are used to characterise the state of

TABLE 3.1
FLEXIBILITY OF PREPREG AND ITS EFFECT ON VARIED PROPERTIES

Material	*Properties*	*Flexibility of prepreg*	
		Soft	*Hard*
Resin	Pre-curing	Low	High
	Consistency	Sticky	Brittle
	Reaction rate	High	Low
Prepreg	*B*-time	High	Low
	Compressibility	High	Low
	Initial softening temperature	Low	High
	Glass transition temperature	Low	High
Laminate	Flow	High	Low
	Scrap	High	Low
	Resin content	Low	High
	Thickness	Low	High
	Structure	Weave exposure	Voids

pre-curing:

(g) *Tack-time*, tested by contacting a prepreg sample to a heated test plate of constant surface temperature.[21,31,36]
(h) *Extent of gelation.*[12]
(i) *Viscometric properties.*[31,37–39]

As to practicability there are some limitations. For testing the *B*-time some resin particles have to be eliminated by 'wrinkling' the prepreg. This results in difficulties when a prepreg is too soft. Then the tack-time is to be tested. The Kofler-method, which has the advantage of rapidity, requires prepregs with glittering surfaces. This is usually realised with fabrics as reinforcements, but not with non-wovens.

For SMC special moulding tools have been developed to characterise the behaviour of compounds relating to flow and mould filling.[40]

3.3.3.2. Behaviour in the Press

The deformation of a prepreg stack under load and temperature as applied in the press can be measured by using sensors for measuring deformation, load, and temperature as a function of time,[35] see Fig. 3.13.

The temperature curve steadily increases up to the curing temperature. At the beginning contact pressure is applied and during heating up it is increased to laminating pressure. The deformation demonstrated in this diagram relates to prepregs of a medium state of pre-curing. This curve is discontinuous, and four striking points

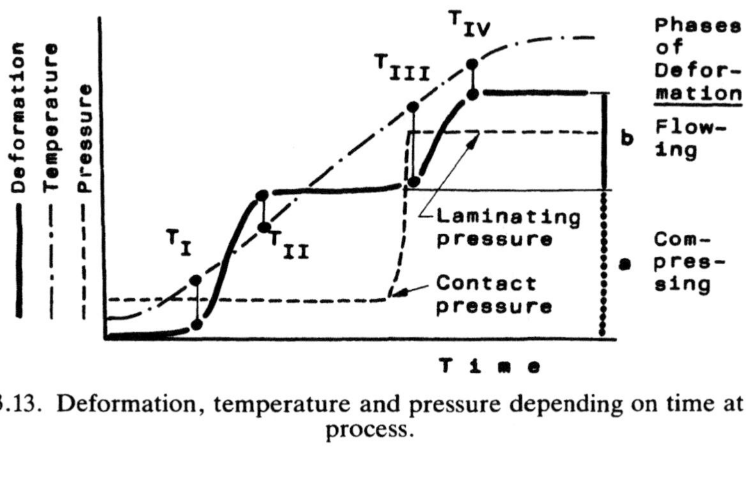

FIG. 3.13. Deformation, temperature and pressure depending on time at press process.

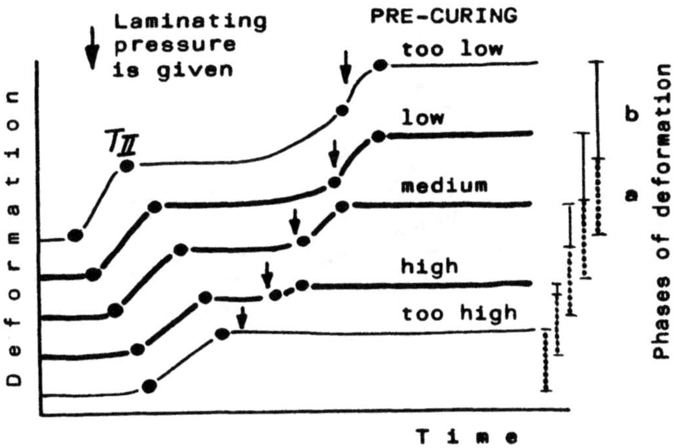

FIG. 3.14. Typical deformation curves originating from prepregs with varying states of pre-curing.

corresponding to specific temperatures can be distinguished as follows:

- T_I beginning to compress the prepreg stack;
- T_{II} end of compression, and beginning of softening the resin;
- T_{III} beginning of increasing flow, caused by the laminating pressure; and
- T_{IV} end of flow.

The deformation can be subdivided into two phases: (a) compression (no scrap); and (b) flow (scrap occurs).

Prepregs with varying states of pre-curing result in characteristic deformation curves, see Fig. 3.14. The lower the pre-curing, the longer the flow phase (resulting in more scrap) and the later laminate pressure can be applied. If pre-curing is too high, no flow can be registered; that means laminates are pressed without any scrap, but they may contain some voids.

For producing laminates with satisfactory surfaces, without any voids, a certain amount of pre-curing is desirable, resulting in a flow of about 10 to 20%. When comparing different resin types, prepregs with specific ranges for the *B*-time are required. On the other hand, measuring the initial softening temperatures resulted in intersecting ranges, see Fig. 3.15. Therefore the latter value is more suitable for predicting the behaviour in the press than the prepreg *B*-time.

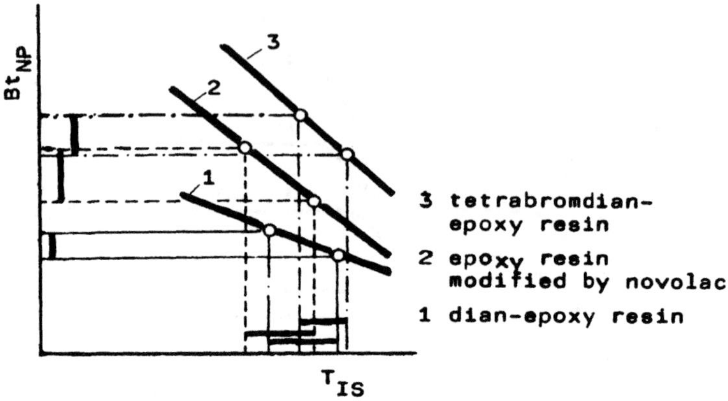

FIG. 3.15. Fairly typical behaviour of different resin types: specific ranges for the prepreg B-time $B_{t_{NP}}$ are required, whereas the analogous ranges for the initial softening temperature T_{IS} are intersecting.

The temperature T_{II} (see Fig. 3.14) which increases the more the resin is pre-cured, nearly corresponds to T_{IS} and T_g. Further interpretations resulted in a correlation of flow to 'relative B-time reduction', z, given by eqn (25), which is to be expected as z depends on the curing rate of prepreg α_P, eqn (26), whereas α_B is to be regarded as constant.

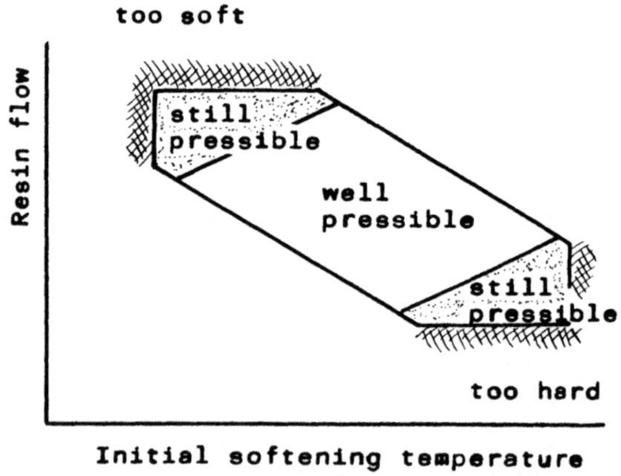

FIG. 3.16. Example for estimation of 'prepreg pressibility' depending on initial softening temperature and resin flow.

Thus, these considerations have resulted in typical diagrams for representing the 'pressibility' of prepregs, as demonstrated in Fig. 3.16. Here the axes are flow and initial softening temperature. Analogously, a diagram like that can be developed depending on z.

3.3.3.3. Effect of Prepreg on Final Properties

The relation between the properties of prepreg materials and those of the ultimate product, i.e. the laminates, has to be understood.[41–45] Important final properties depend on the application[20,22,23] but may be the mechanical, electrical, dielectric, chemical (i.e. effect of water and common solvents), or thermal (e.g. the T_g).

Naturally these properties depend very much on the reinforcement and matrix selected. But also, for a given set of starting materials, they depend on the prepreg process. Using thermo-analytical methods such as DSC,[13,34] for instance, it is possible to optimise the whole process and achieve the maximum value for the laminate T_g.

The use of fundamental methods of this kind is possible only where all interactions are sufficiently known and there is enough background knowledge. However, if it is impossible to apply fundamental scientific techniques there are still many ingenious empirical methods which can be used.

3.3.3.4. Mathematical Simulations

Recently, trial procedures have become available for making fundamental approaches to the above mentioned problems, starting with mathematical simulations for analysing the whole process.

The simulation suggested by Halpin *et al.*[46] (see Fig. 3.17) is supplemented by physical, chemical and technological functions for each case. Based on resin properties, the processes of cure and flow and their effect on laminate quality, especially void content, are discussed. This kind of procedure helps further developments to be understood.

There are also proposals by Loos and Springer,[47] who have suggested particular mathematical simulations relating to cure, flow, void formation and stresses within the laminate. The various parameters can, in principle, be followed by computer-aided investigations during cure. Problems can arise, however, if some of the property data necessary for calculation are not available.

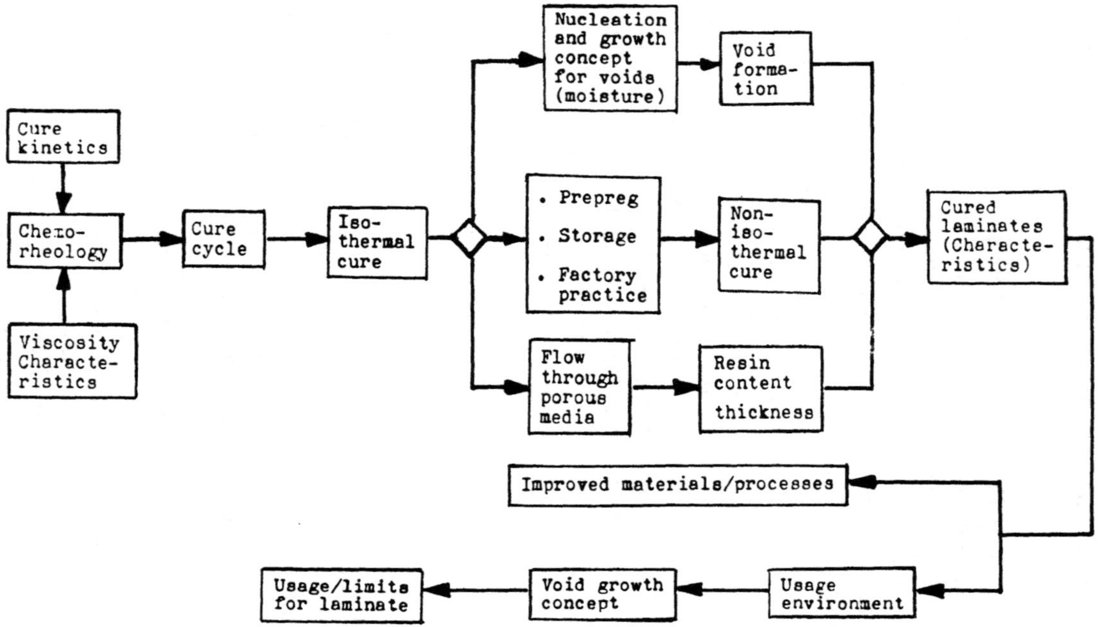

FIG. 3.17. Survey on mathematical simulations of a total polymer process based on prepregs according to Halpin *et al.*[46]

3.4. TECHNOLOGY OF THE PREPREG PROCESS

Prepregs of high quality, based on fabrics, are preferably produced on vertical plants. The term 'vertical' relates to the dryer which is the core of such an impregnating plant. In general it consists of building units as follows (see Fig. 3.1).

(a) *Unwind unit*: This consists mainly of a double let-off, web-storing unit with bonding press or sewing machine, broad stretching roller, exhaust unit and heating roll for heating up the web.
(b) *Impregnating unit* (wet end): Consisting of dipping rollers and an impregnating bath partly fitted with controlled heating.
(c) *Vertical dryer*: Mostly built up with two tunnels (double dryer) to evaporate the solvents and to partly crosslink the polymer (pre-cure the resin).
(d) *Roll up*: Mostly with double take-off alternative to *cross cutter* type guillotine, rotational system or realised by laser.

In general, prepreg plants are equipped with beta-gauges for measurement of fabric and prepreg weight to control the pick-up of resin, and

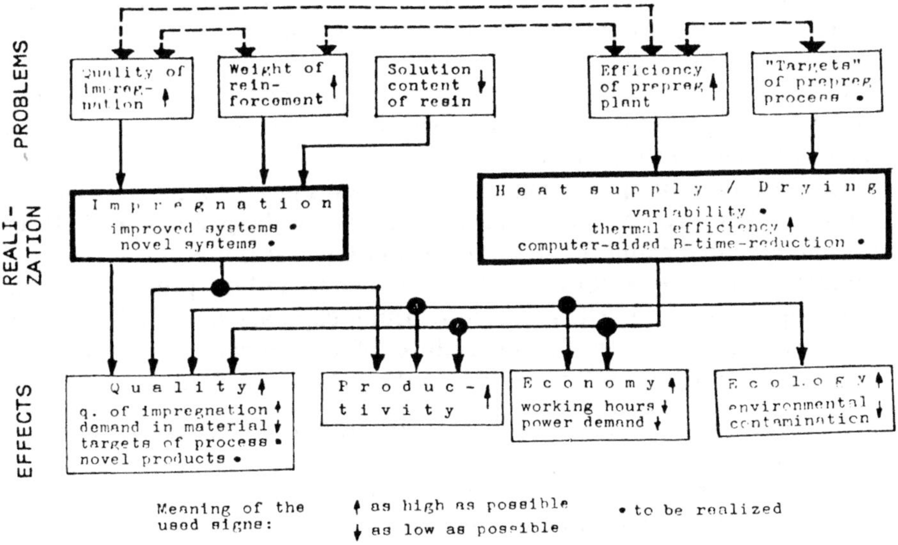

FIG. 3.18. Problems and effects of the prepreg process originated from desired developments with conclusions as to their technological realisation.

the viscosity of the resin mixture, also called 'varnish', is controlled too. Controlling the temperature within the dryer tunnels will be discussed in Section 3.4.2.

These principal characteristics are achieved in different ways by different engineers, so for any details, the reader is referred to technical literature from individual companies.

There are two main problems in this technology (see Fig. 3.18)—the impregnation, and the heating of the dryer.

3.4.1. Impregnation

Consider a vertical plant operating with dip impregnation. The procedure for 'prepreg forming' depends mainly on two functions as follows (see Fig. 3.19(a)).

(a) The *prepreg weight,* m_P, corresponding to the web speed v with varnish viscosity η as parameter.
(b) The *prepreg B-time,* B_{t_P}, depending on the web speed v, and with temperatures T of the dryer as parameters.

Starting with m_P as a 'target' we find a certain web speed when choosing a viscosity. Then, starting with target B_{t_P}, such a temperature T has to be adjusted to correspond to the given web speed v. So, finding the parameters (η, v, T) to meet the 'targets' (m_P, B_{t_P}) requires an iterative approach.

But, in practice, the problems are more complicated because fabrics differ in their ability to take up varnish and different resin types may have different wetting properties.

Considering the problems presented in Fig. 3.18 fabrics with a higher weight are desirable and an increased web speed as well. Both the parameters result in a better economy but a poorer quality of impregnation (see Section 3.2.1). In order to reduce prepreg defects and imperfections, particularly concerning capillary lines, efforts have to be concentrated on the procedure of impregnation. Conceivable and actual procedures[48–52] are directed to physical effects as follows:

(a) Removal of *absorbed coats* on the reinforcement (by pre-drying).
(b) Increase of the *surface energy* of the reinforcement (by electrostatic treatment, plasma).
(c) Decrease of *surface tension* of the resin mixture (by special resins and solvents).

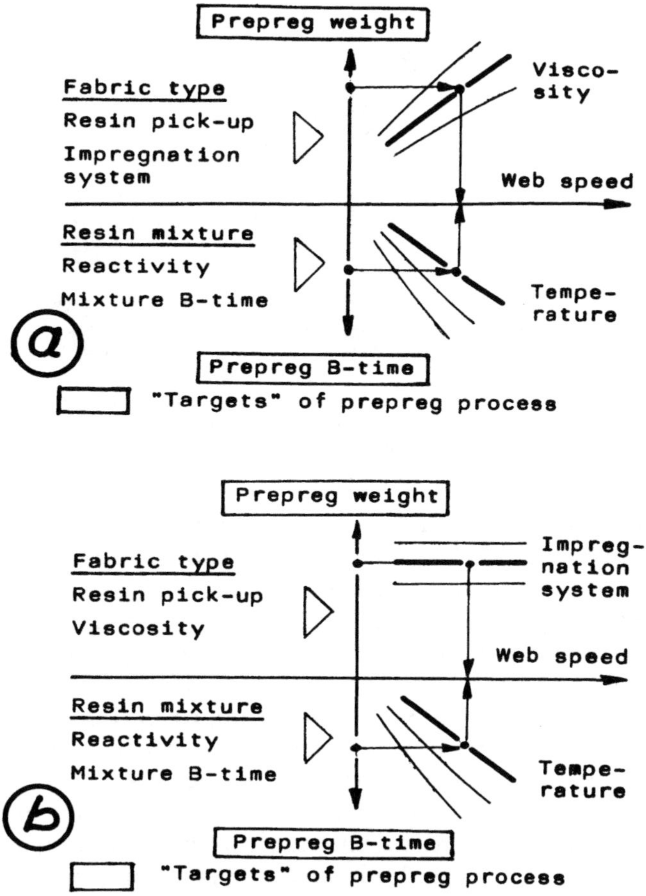

FIG. 3.19. Typical diagrams of prepreg process—influences and 'targets'. (a) Classical impregnation (dipping); (b) improved impregnation (comb-like doctor blade).

(d) Improving the *liquid-flow through the capillary system* of reinforcement (by mechanical oscillators or ultrasonics by using several impregnating baths, or larger ones, by having multiple impregnations with 'breathing rolls' or specially designed doctor blades, pre-wetting rolls, or vacuum devices).

We have achieved satisfactory results by using a 'repeated impregnating system'[53] (see Fig. 3.20). The first impregnation is done by a

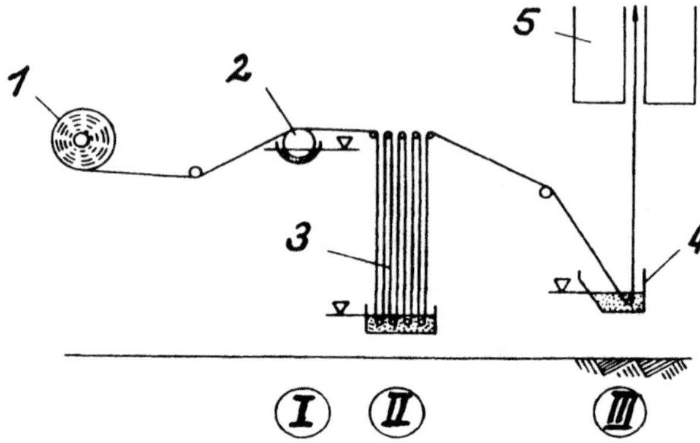

FIG. 3.20. 'Repeated impregnating system' to improve the quality of impregnation. I, Pre-wetting; II, extending of impregnation time, and massage; III, dipping. 1, Fabric roll; 2, pre-wetting roll; 3, impregnating storage unit; 4, impregnating bath; 5, dryer.

pre-wetting roll so that the fabric is penetrated by the varnish from one side, whereas air is enabled to escape to the opposite side. Then the web, being wet, runs through an impregnating storage unit, extending the impregnating time and effecting a certain massage. After this pre-treatment the web is dipped into the impregnating bath and the varnish pick-up is carried out. In this way the quality of impregnation (see Section 3.2.1) can be improved from mark 2, and 2 to 3, when simple dipping was employed, down to mark 0 to 1. The prepregs resulted in laminates without any defects.

Figure 3.19(a) shows a diagram of the classical impregnation process. It is complicated to find both targets 'prepreg weight' and 'prepreg B-time' because both characteristics depend on the web speed, i.e. they cannot be chosen freely. In each case the viscosity or the temperature have also to be changed. The improved impregnation system, see Fig. 3.19(b), enables the operator to find the process parameters easily, because the web speed can be changed for adjusting the prepreg B-time without any influence on the prepreg weight. Furthermore, this system makes it possible to apply varnishes with a considerably reduced solvent content. Thus, the environment is less polluted by solvent vapour—assuming that no expensive equipment for combustion or recovery are available—and the capacity of the prepreg plant can be increased because less solvent bulk has to be evaporated.

A variant of this impregnating principle is shown in Fig. 3.21.[54] By

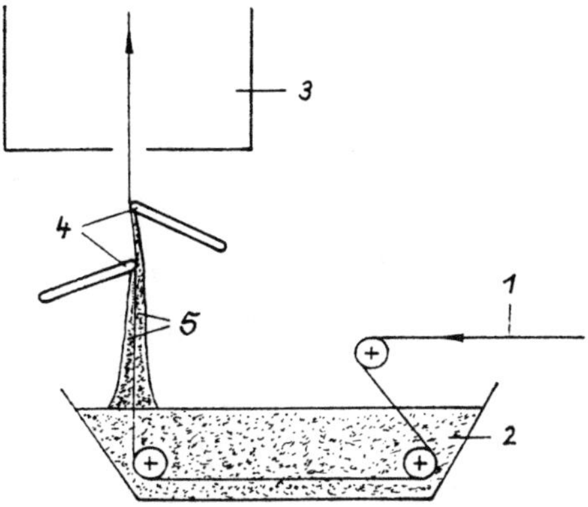

FIG. 3.21. Variant for an impregnation principle using comb-like doctor blades resulting in an independence of prepreg weight on web speed (see Fig. 8.19(b)). 1, Web; 2, impregnating bath; 3, dryer; 4, comb-like doctor blade; 5, resin flowing down.

using specially designed doctor blades (comb-like), arranged at both sides of the web, only enough varnish is allowed to pass through to give the correct prepreg weight. Independence of viscosity is based on the fact that fluctuations in solid content are much smaller than those in viscosity.

3.4.2. Heat Supply—Drying

Vertical dryers only need a small area but a large height. The heat is applied in an ascending and a descending branch resulting in a temperature profile showing a typical double-humpy curve which can be subdivided into two zones connected by a transitional zone for diverting the web. Six different periods (see Fig. 3.22) can be distinguished as follows:

Zone I:
(1) Heating up of the web
(2) Evaporating of the solvents as even as possible
(3) First phase of pre-curing

Transitional zone:
(4) Freezing of polymerisation, resulting in a tack-free surface

Zone II:
(5) Second phase of pre-curing
(6) Cooling down of the web (prepreg) at the exit from the tunnel.

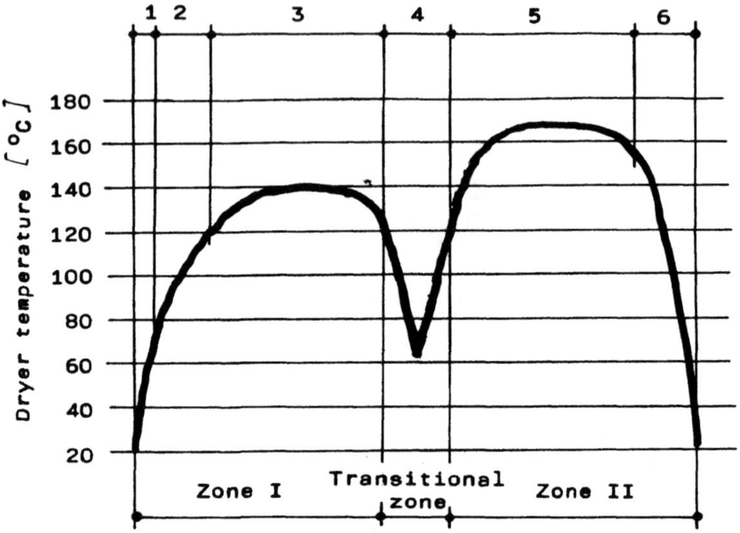

FIG. 3.22. Fairly typical double-humpy curve showing a temperature profile of a double-dryer.

The events mentioned above do not strictly run one after the other, but there are transitions. After entering the dryer the temperature should be kept below the boiling-point of the solvent and then slowly be increased. If the web has heated up to such a temperature that evaporation begins some of the solvent will already have escaped from the surface coat. Besides, such a low viscosity is needed to release the solvents easily from the web core, avoiding the formation of coatings, craters, and voids.

Production plants should provide for flexibility in the adaption of temperature profile used with various solvents and resins. The profile in question should be controlled constantly to $\pm 1{\cdot}5$ K.

At present, production plants operate by convection drying (see Fig. 3.23(a)) using an indirect heat supply by steam, hot water, or hot oil. Hot air, supplied by post-combustion equipment, is also used. Generally the dryers are arranged to have a counter current with a by-pass, and the incoming air is blown through nozzles against the web in order to improve the efficiency of the dryer by dissolving of the surface coat. But altogether the thermal efficiency of a convection dryer for this purpose is poor because a large bulk of air has to be

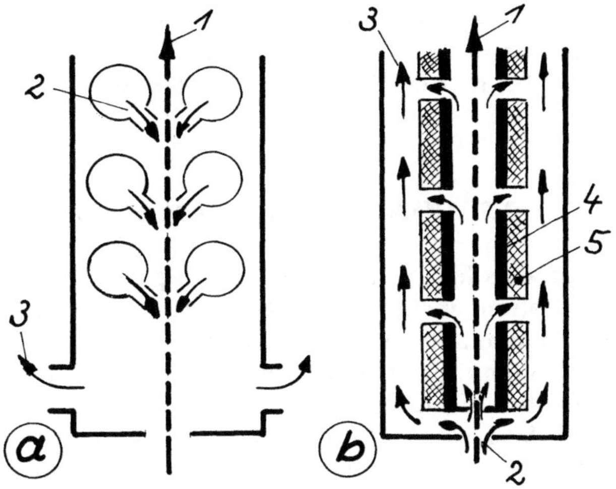

FIG. 3.23. Principles of vertical dryers. (a) Pure convection system; (b) combined system (thermal radiation and convection). 1, Web; 2, incoming air; 3, outgoing air; 4, heating plate; 5, thermal insulation.

heated up to avoid risks of explosion, i.e. to decrease the solvent: air ratio.

Combinations of convection and thermal radiation have also become known.[55] These systems operate with a higher thermal efficiency because cold air is used for diluting solvent vapour. By employing radiation heating plates, indirectly heated with hot oil, the temperature profile can be varied over a wide range, see Fig. 3.23(b). Because the air is removed there always exists an underpressure within the dryer preventing the escape of solvent vapour into the workshop.

At the web entry the air flow is subdivided into a heavy stream, remaining cold to dilute the solvent vapour, and a small stream getting into the inner part of the dryer, to be heated up at the heating plates. Thus, the web is heated up both by thermal radiation and convection.

The temperature profile is measured by sensors (thermocouples) arranged along the web run and recorded. The radiation plates are connected with different heating circuits which are controlled. Thus, a desirable *B*-time reduction or curing rate can be achieved, preferably in a computer-aided manner. *B*-times tested in practice have agreed well with values calculated by the equation given in Sections 3.2.3.2 and 3.2.3.3.

A combination dryer like that provides a specific web speed of 0·4 to 0·5 m/min per metre length of the dryer, and the specific energy supply is about 1·5 to 2·0 kWh/kg prepreg (resin content about 40 to 50%).

3.5. CONCLUDING REMARKS

The prepreg process has been discussed as a part of the total polymer processing operation, including all the stages of chemical conversion and material shaping. Its core is the concept of partial crosslinking, involving calculations for *B*-time reduction and curing rate based on resin characteristics. Thus the forecasting of technological parameters is possible and the forecasts have proved by experience to be correct. Possibilities for the computer-aided control of the temperature profile of the dryer have also been mentioned. Testing of prepregs is particularly significant with regard to their further processing, especially for their behaviour in the press. Discussion of the technology of the prepreg process, towards the end of this chapter, have been concentrated on the problems of impregnation and the heating of the dryer.

After the remarkable progress in this field in recent years future developments will relate to the application of scientific knowledge to normal production procedures.

As to the reinforcements, combined structures will be more widely used—hybrids of advanced and conventional textile fibres, manufactured both traditionally and by use of novel technologies, e.g. stitch-bonding techniques and tri-axial weaving. In this way, novel designs for reinforced thermosetting and thermoplastic polymers will contribute to the solution of more problems.

Concerning matrices, it can be expected that improved and novel resins will be able to meet the higher requirements demanded of them, e.g. higher thermal stability, better chemical resistance, curing actions which are always reproducible, better shelf-life, and solvent-free formulations.

In the near future, scientific test methods, e.g. FT–IR, DSC, DMA, and HPLC will not only be used for the characterisation of materials to check their identification, but in all probability to contribute directly to the running of prepreg processes in order to predict the behaviour of prepregs during processing in a comprehensive manner.

In future, not only the optimisation of process actions but also their mathematical simulation will be employed increasingly, especially through computer-aided methods. Quality control will be integrated into this process, and process control will be further improved.

The prepreg process, in general, will continue to be used. That means, the possibility of its substitution by a total novel technique is not envisaged at present. But, of course, innovations will be made relating to details of materials and plant, and to process engineering as well. This statement is valid for the prepreg process and also for further processing up to the stage of shaping composite products, sometimes in conjunction with filament winding or pultrusion. Thus, double pressure-belt plants to press large quantities of flat laminates continuously will increasingly come into use.

The differences between thermosetting and thermoplastic polymers will progressively disappear—thus, the 'thermoplastic prepreg' is estimated to have a good chance, as well as crosslinkable prepregs.

The employment of robots ought to be mentioned, e.g. for handling prepreg sheets. And, last, but not least, the combination of SMC with 'advanced prepregs' is expected to result in many advantages with respect to design, processing, and applications.

REFERENCES

1. Newman, D. J., in *Developments in Reinforced Plastics—2,* ed. G. Pritchard, Applied Science Publishers Ltd, London, Chapter 7, 1982.
2. Schriever, H., Seiler, E. and Welz, M., *16th AVK Reinforced Plastics Conf.* Freudenstadt, FRG, Paper 8a, 1980.
3. Derek, H., The technology of processing sheet molding compound, Thesis, Rhein.-Westfälische Technische Hochschule Fakultät für Maschinenwesen, Aachen, 1982.
4. Schwartz, R. I. and Schwartz, H. S. (eds), *Fundamental Aspects of Fiber Reinforced Plastic Composites,* Interscience Publishers, New York, 1968.
5. Pritchard, G. (ed.), in *Developments in Reinforced Plastics—2,* Applied Science Publishers Ltd, London, Chapter 2, 1982.
6. Springer, G. S., in *Developments in Reinforced Plastics—2,* ed. G. Pritchard, Applied Science Publishers Ltd, London, Chapter 3, 1983.
7. IPC-A-600 B, *Acceptability of Printed Wiring Boards,* Inst. of Printed Circuits, USA.
8. Wiedemann, G., Rothe, H., Haase, A. and Wiechmann, R., *Plastverarbeiter,* **32** (4), 459 (1981).
9. Wiedemann, G., Rothe, H., Keusch, S. and Haase, A. *Plastverarbeiter,* **33** (5), 572 (1982).

10. GILLHAM, J. K., in *Developments in Polymer Characterisation*—3, ed. J. V. Dawkins, Applied Science Publishers Ltd, London, 1982.
11. ENNS, I. B. and GILLHAM, J. K., *J. of Appl. Pol. Sci.*, **28,** 2831 (1983).
12. HAGNAUER, G. L. and DUNN, D. A., *Ind. Eng. Chem. Proc. Res. Dev.*, **21** (1), 68 (1982).
13. CRAVEN, P. M. and BRAMEL, J. B., *Proceedings of the Printed Circuit World Convention II,* Vol. II, E. G. Leuze Verlag, Saulgau/Württ., p. 27, 1981.
14. PERSSON, K. E., *Printed Circuit Meeting,* Milan, 1967.
15. ASTM D 1200, American Society for Testing and Materials, 1916 Race St, Philadelphia, USA.
16. *Manual*: *Heating Bench 'System Kofler',* C. Reichert AG, Hernalser Hauptstrasse 219, Vienna.
17. MIL-G-55 636, Military Specification, US Army, Navy and Air Force.
18. LN 65 090, FRG air craft standard, Beuth-Vertrieb, Cologne, Germany.
19. LN 29 530, FRG air craft standard, Beuth-Vertrieb, Cologne, Germany.
20. DIN 16 913, Part 1, 2, 3, German Industry Standard, Beuth-Vertrieb, Cologne, Germany.
21. Standard Specification IPC-L-110 A, Institute of Printed Circuits, 1717 Howard St, Evanston, Illinois, USA, 1970.
22. Standard Specification IPC-L-109 A, Institute of Printed Circuits, 1717 Howard St, Evanston, Illinois, USA, 1980.
23. Test Methods Manual IPC-TM-650, Institute of Printed Circuits, 1717 Howard St, Evanston, Illinois, USA.
24. SENTURIA, S. D., SHEPPARD, N. F., LEE, L. and DAY, D. R., *J. Adhesion,* **15** (1), 69 (1982).
25. SANJANA, Z. N. and RAY, I. R., *Amer. Chem. Soc. Div. Polym. Chem. Polym. Progr.*, **22** (2), 255 (1981).
26. MENGES, G., HILLE, E. A. and THIELE, W., *Kunststoffe,* **69** (8), 467 (1979).
27. ASTM D 635-56 T, American Society for Testing and Materials, 1916 Race St, Philadelphia, USA.
28. WIEDEMANN, G., *Elektrie,* **35** (8), 428 (1981).
29. WIEDEMANN, G., *Plaste u. Kautschuk,* **30** (7), 369 (1983).
30. TGL 35 077/01, Standard of the GDR, Publishers for Standardisation, Export for Standards, DDR-7010 Leipzig, POB 1068.
31. WITSCHAS, H., *Plastverarbeiter,* **29** (5), 253 (1978).
32. *Manual: Micro Heating Table 'System Boëtius',* VEB Analytik, DDR-8019 Dresden, Augsburger Strasse 57; Export by: Intermed-Export-Import, Schicklerstrasse 5/7, DDR-1020, Berlin, POB 17/GDR.
33. WIECHMANN, R. and ZEPNIK, S., *Plastverarbeiter,* **30** (9), 519 (1979).
34. CUMMINGS, L. C., *37th SPI Reinforced Plastics/Composites Conf.*, USA, Paper 29 E, 1982.
35. WIEDEMANN, G., STEPHAN, M., ROTHE, H. and GARSKE, B., *Plaste und Kautschuk,* **30** (6), 332 (1983).
36. US Patent 4 312 212, US Class. 73/15.4, 1982.
37. BLOECHLE, D. P., *J. of Elastomers and Plastics,* **10** (4), 377 (1978).
38. ROLLER, M. B., *Polym. Eng. Sci.*, **15** (6), 406 (1975).

39. TAJIMA, Y. A., *Polym. Comp.*, **3** (3), 162 (1982).
40. FOCK, K. and KITTL, R., *15th AVK Reinforced Plastics Conf.* Freudenstadt, FRG, Paper 11, 1978.
41. Standard Specification IPC-L-108, 115, Institute of Printed Circuits, 1717 Howard St, Evanston, Illinois, USA.
42. DIN Draft IEC 52 (CO) 212, German Industry Standard, Beuth-Vertrieb, Cologne, Germany.
43. DIN Draft 40 802, German Industry Standard, Beuth-Vertrieb, Cologne, Germany.
44. TGL 11 651. Standard for the GDR, Publishers for Standardisation, Export for Standards, DDR-7010 Leipzig, POB 1068.
45. TGL 25 016. Standard for the GDR, Publishers for Standardisation, Export for Standards, DDR-7010 Leipzig, POB 1068.
46. HALPIN, I. C., KARDOS, I. L. and DUDUKOVIC, M. P., *Pure and Applied Chem.*, **55** (5), 893 (1983).
47. LOOS, A. C. and SPRINGER, G. S., *J. of Comp. Mat.*, **17** (3), 135 (1983).
48. Caratsch AG, Microelectronik Conf., Moscow, p. 152, 1976.
49. BLALOCK, P. and FRASER, R., *Periodical of the Burlington Glass Fabrics Co.*, Burlington Ind., New York, Vol. 4, 1972.
50. SARAZIN, R. G. and PAULUS, I. R., *World Conference Electr.*, Moscow, Sect. 3-A, Paper 40, 1977.
51. VITZ, W., *World Conference Electr.*, Moscow, Sect. 3-A, Paper 56, 1977.
52. JUNGE, K. H., *Plaste und Kautschuk*, **27** (3), 157 (1980).
53. GDR Patent DD-PS 148 628, Class. C08J, 1983.
54. GDR Patent DD-PS 209 146, Class. B29D, 1984.
55. GDR Patent DD-PS 212 574, Class. F26B, 1984.

Chapter 4

CURING PROCESSES OF ACETYLENE TERMINATED RESINS

CHARLES Y-C. LEE
Air Force Wright Aeronautical Laboratories, Ohio, USA

SUMMARY

Acetylene terminated resins are a new class of thermoset resin, which can be cured through a thermally induced, radical initiated addition reaction. The combined features of curing without evolution of volatiles, good thermal stability and moisture resistance of the reaction products make the acetylene terminated resin technology an attractive means to achieve high temperature matrices for advanced composites. This chapter will discuss the curing characteristics of these resins. The thermal characteristics of the curing reaction are rather insensitive to the structures of the resins, but the processing of these resins is highly dependent upon the initial T_g of the resins. The discussion is separated into the processing of the low initial T_g resins and that of the high initial T_g ones. The effects of cure conditions and extent of cure on the properties of the cured matrices are also discussed.

4.1. INTRODUCTION

In the past decade there have been substantial research activities in the area of acetylene containing prepolymers as thermoset resins for achieving high temperature matrices for structural applications.[1] The cure reaction of these prepolymers can be induced thermally. The cure mechanism is through the addition reaction of the acetylene, thus presenting a desirable advantage in the processing of complex parts, since no volatile by-products will be evolved during the processing. While the properties and thermal stability of these acetylene containing resins are highly dependent on the chemical structures of the remaining molecules, the cure reaction products of the acetylene groups are thermally stable and do not show decrease in high temperature performance properties due to absorption of moisture, or in the jargon of the structural resin technology, do not have the 'hot–wet' problem as found in epoxies.

A recent article by Hergenrother[2] gave an excellent overview of the synthesis of various acetylene containing prepolymers and their properties. These prepolymers include polymer chains terminated at both ends, or only at one end, with acetylene groups, chains with pendant acetylene, ester oligomers (ethynyl-phenyl group connected to an ester), propargyl containing polymers, and polymers containing diacetylene, enyne, or phenyl substituted acetylene.

This chapter will discuss only the curing characteristics and the processing of the prepolymers that are terminated at both ends with ethynyl-phenyl groups. These resins are referred to as the acetylene (AT) resins. The intrinsic nature of the thermally induced acetylene reaction of these polymers is rather insensitive to the variations of the structures of the remaining chains, so the discussions in the rest of this chapter, instead of giving an account of the processing characteristics of each resin, will be as general as possible.

4.2. CLASSIFICATION OF AT RESINS

For processing considerations, these acetylene terminated (AT) resins can be classified into two categories according to their initial glass transition temperature (T_g) before curing: high and low initial T_g resins. The processing of the 'low initial T_g' AT resins, defined as those with initial T_g below 100°C, can be quite different from those with initial T_g above 100°C, classified here as the 'high initial T_g' AT resins. This classification will be used throughout, with the exception of paragraphs concerning synthetic methods.

An alternative classification of the AT resins is based on the chemical structures of the molecules[3] and relates to the expected thermal stability of these structures. The high thermal stability AT resins are composed mainly of aromatic, heterocyclic structures. The molecular motions of these structures are rigid and resins based on these structures usually have high initial T_g. The lower stability AT resins usually have more flexible ether linkages in the chain, thus the initial T_g values are usually low. It should be emphasised that the term 'lower stability' is only relative when compared with the aromatic heterocyclic structures. They are all intended for high temperature applications to at least 350°F (177°C).

Generally speaking, for most, but not all, of the AT resins studied thus far, these two classifications are very similar: a high stability resin is also a high initial T_g resin. It must be recognised, however, that the initial T_g of an AT system based on a certain chemical moiety can change with the length of the chain. Thus it is possible that a high stability AT system will have to be processed like a low initial T_g system, and vice versa.

Most of the high thermal stability AT resins are synthesised with ethynyl-containing precursors, which also contain the necessary moieties to react with other monomers to form the desired heterocyclic aromatic structures. Such precursors include the ethynyl-containing amines, diamines, and benzils. The resins in the high thermal stability category include the benzimidazoles,[4] the phenylenes[5] ('H-resins'), the imides,[6–10] the isoimides[11] and the quinoxalines.[12–16]

The low stability AT resins, referred to as the phenylene-R series,[17] have been synthesised both with an ethynyl-containing precursor (a metal alkynide) to form the ether linkage of the chain, and with polymer chains terminated with suitable replacement groups, which are subsequently replaced by acetylene using appropriate reactions. The resins in this category include those based on thiodiphenol (ATT),[15] resorcinol (ATP),[15] sulphonyldiphenol (ATS),[16,17] and isopropyldiphenol (bis-phenol-A) (ATB).[18]

Most of the research activities in the AT resin area are concentrated on producing properties suitable for structural applications; thus the backbone structures are usually similar to those which, in high molecular weight linear chains, yield high temperature thermoplastics. It should be pointed out, as an illustration of the versatility of the AT technology, that the synthesis of AT perfluoroalkylene ether benoxazole had been reported,[19] and the cured polymer was a thermally stable rubbery material.

This section is a very brief, over-simplified description of the synthesis of AT resins. Readers interested in the synthesis of these resins are referred to the previously mentioned treatise by Hergenrother[2] and the references contained therein.

4.3. CURE KINETICS

4.3.1. Activation Energies

Cure kinetic parameters of thermoset resins can be obtained by several approaches. An estimation of the activation energy can be obtained by plotting the log of the gelation times from isothermal cure experiments versus the reciprocal of the experimental temperatures.[20] The slope, n, of such a plot is related to the activation energy of the reaction, E, by the equation

$$n = E/(2{\cdot}303R) \quad (1)$$

where R is the gas constant. This equation is derived on the assumption that the extent of cure at gelation is a constant and is not influenced by the experimental temperature.

Experimentally, gelation times of thermoset resins were often measured by such techniques as torsional braid analysis (TBA),[21] torsion impregnated cloth analysis (TICA),[22] or dynamic viscosity measurement.[23] All these measurements, however, tend to overestimate the activation energy because the measured 'gelation' event actually occurs at a higher extent of cure with increasing experimental temperature. Also, the calculated value is meaningful only if the 'gelation' event is not affected by the vitrification process caused by the increase of the glass transition temperature (T_g) during the isothermal cure. Thus, these methods are only valid for AT resins with low initial T_g.

In spite of their shortcomings, these measurements offer practical information for processing considerations and can provide a convenient means to estimate the activation energy of the cure kinetics of thermoset resins.

Another common method of obtaining kinetic information is to analyse the exotherm profiles of thermoset resins obtained from differential scanning calorimetry (DSC) experiments.[24,25] One of these approaches,[25] extensively used in the Materials Laboratory to analyse the kinetics of AT resins, involves digitising the exotherm profiles obtained with different temperature scanning rates, and analysing these data with a computer program to calculate the activation energy at different extents of cure. If there is only one rate-determining step in the cure reaction the calculated activation energy should be independent of the extent of cure.

Table 4.1 lists the activation energies of various AT resins obtained

TABLE 4.1
ACTIVATION ENERGIES OF AT RESINS

Resin designation	*Chemical type*	*E*	
		kcal/mol	*kJ/mol*
BATQ-H(G)	quinoxaline	20·6	86·1
BATQ-H($n = 0$)	quinoxaline	19·8	82·8
SBQ	quinoxaline	23·8	99·5
ATS(G)	sulphonyldiphenol	21·6	90·3
ATB	isopropyldiphenol	23·9	99·9
ATT	thiodiphenol	24·7	103·2

(G) Synthesised by Gulf Research Corp.

from the DSC experiments.[26] The range of the activation energy appears to be between 20 to 25 kcal/mol (80 to 100 kJ/mol). The chemical structures of the molecular chain between the acetylene end groups do not appear to influence the kinetics significantly. Based on the data from purified monomers, the heat of reaction of the cure is estimated to be 60·3 kcal/mol (252 kJ/mol) of AT resin (or 30·1 kcal/mol (526 kJ/mol) of acetylene).

4.3.2. Reaction Exotherm

Figure 4.1 shows the DSC trace of an AT resin with low initial T_g.[27] This resin, with a chemical structure based on the sulphonyldiphenol, has an initial T_g of about 35°C. This exotherm profile is typical of a low initial T_g AT resin. The onset of the exotherm, obtained with a scanning rate of 10°C/min, typically is at around 140°C, and peaks at about 230°C.

For a high initial T_g AT resin the activation energy of the chemical reaction itself is not significantly different, but the reaction rate is noticeably influenced by the initial T_g of the resin. Figure 4.2 shows[28] the DSC trace of an AT resin with a quinoxaline structure. The onset of the exotherm, unlike the smooth onset shown in Fig. 4.1, is abrupt, and occurs right at the initial glass transition temperature of the resin.

This abrupt change in the exotherm profile is due to the influence on the kinetic rate of the restricted mobility of the molecules below the glass transition temperature. Even though the molecules have sufficient thermal energy to react, the restricted mobility reduces the encounter frequency of the reactive groups, causing a decrease in the

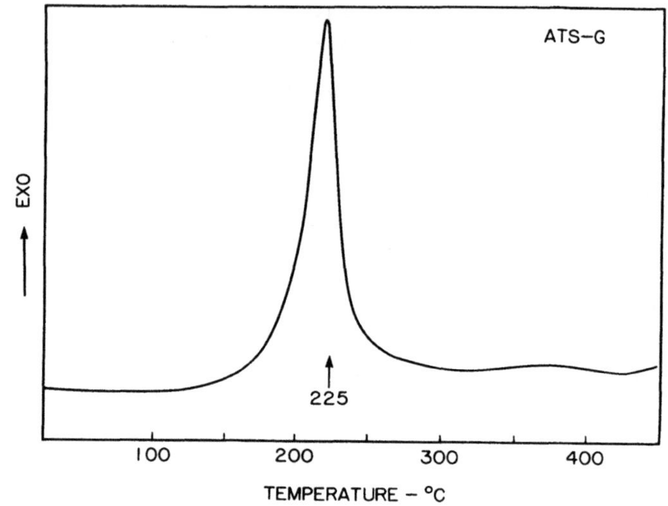

FIG. 4.1. DSC trace of a typical low initial T_g AT resin (ATS).

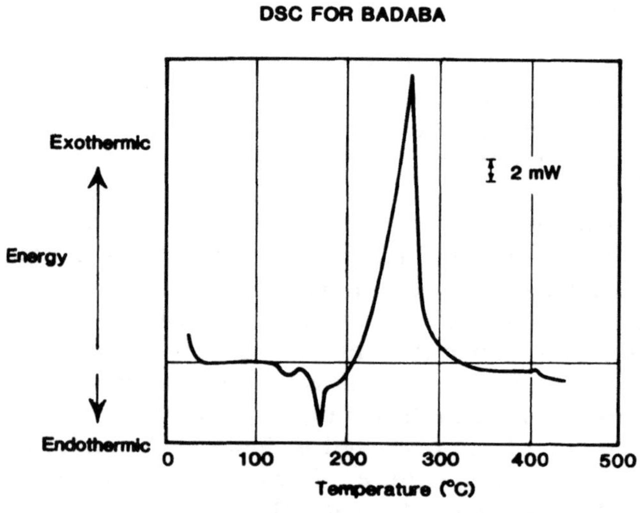

FIG. 4.2. DSC trace of a typical high initial T_g AT resin (BATQ, $n = 0$).

apparent kinetic rate. This is analogous to the 'diffusion controlled' quenching of thermoset reactions,[29] the effect of which is the prevention of complete cure if the cure temperature is significantly below the T_g of the fully cured resin.

Because of this 'diffusion controlled' effect the kinetic rate constant, instead of following an increase with temperature in accordance with the Arrhenius equation, will increase abruptly upon the increased mobility of the molecules when the temperature exceeds the glass transition temperature.

4.4. PROCESSING OF LOW INITIAL T_g RESINS

4.4.1. Viscosity

Processing of thermoset resins is usually carried out above the initial T_g. The processing conditions are highly dependent on the viscosity of the resins. The low initial T_g AT resins usually have a range of temperatures where the viscosity is suitable for various types of processing, and yet the cure will not be substantially advanced within a reasonable time period under isothermal conditions.

The viscosity profiles of several low initial T_g AT resins are shown[18] in Fig. 4.3. Measurements were made with a Rheometrics Mechanical

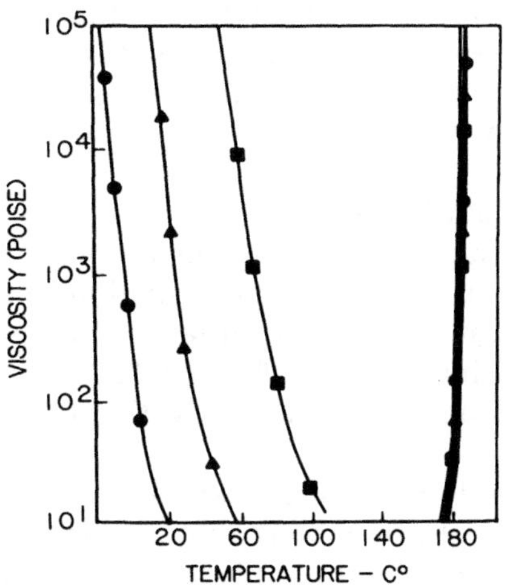

FIG. 4.3. Viscosity profiles of sulphonyldiphenol based (■, ATS), isopropyldiphenol based (▲, ATB) and resorcinol based (●, ATP) AT resins at a temperature scan rate of 2°C/min.

Spectrometer, using the parallel plate method. The scanning rate of these experiments was 2°C/min, and the experimental frequency was 10 rad/s.

The viscosity decreases with increasing temperature at the lower temperature range until the kinetics of cure eventually produces an increase in viscosity. In the lower temperature range where the viscosity decreases with increasing temperature, the viscosity profiles of the resins are very similar in shape, and appear to have shifted in temperature with respect to each other. The order of the shifting is directly related to the initial T_g's of the resins shown: the T_g's of ATP, ATB, and ATS measured by dynamic mechanical methods are −35°C, 0°C and 35°C, respectively.

The viscosity profiles below 10 P are not recorded because they are out of the sensitivity range of the usual experimental set-up. But in one experiment, where the equipment configuration was set up specifically to make measurements in the lower viscosity range, the minimum viscosity of ATS in a 2°C/min scan was found to be about 40 cP and was reached at 155°C. Because of the insensitivity of the kinetic parameters to chemical structure, as discussed in the previous section, it is estimated that the minimum viscosity temperature of other resins under similar scan conditions will have similar values and be reached within ±10°C of that of ATS.

This supposition is further supported by the viscosity profiles of the resins at the high temperature region in Fig. 4.3. Regardless of the chemical structures of the resins all profiles show rapid increases in viscosity at 175°C to 180°C.

4.4.2. Processing Conditions

The first phase of most processing operations usually requires bulk transport of the resins, e.g. fibre impregnation, mould filling, etc. It is usually not a problem for the low initial T_g AT resins to carry out this phase of the processing below 140°C. At 120°C, the resins can usually undergo isothermal hold for 2 to 3 h without significantly increasing the viscosity.

Such, of course, will not be the case if the palladium catalyst used in the synthesis step was incompletely removed from the resin during the purification procedure. All the information on processing AT was obtained from resins with less than 20 ppm residual catalyst.

Processing temperatures of low initial T_g AT resins can thus be adjusted to obtain a suitable viscosity. For example, ATS can be degassed at 120°C. It is melt prepreggable at about 100°C, and forms films at about 85°C.

All the AT resins shown in Fig. 4.3 are melt prepreggable without the aid of solvents. However, only the ATB prepregs will show the room temperature tack and drape characteristics sometimes required in composite processing. The viscosities of ATS and ATP, as indicated in Fig. 4.3, are too high and too low, respectively, but they can be blended to achieve the desired tack and drape properties.[28]

After the initial phase of processing the second step is usually to carry out the cure to beyond the gelation point in moulds, or under pressure, to set permanently the form or shape of the parts. This is followed by a free-standing post-cure to complete the reaction. For AT resins, the second step can normally be carried out in the range between 140°C and 180°C. Although curing at a higher temperature can realise economy in cure time, there is evidence that low temperature cure can improve the mechanical properties of cured AT matrices. This point will be discussed in further detail in a later section.

Another consideration in the selection of the cure temperature is the possibility of void formation due to volatilisation of the resin or any residual volatiles incompletely removed during the degassing cycle. This is especially important for the very low T_g resins, like ATP and ATB, which usually have higher vapour pressures and lower viscosities at the curing temperature, both factors being favourable to void formation. Especially for ATP, a long holding time at a low cure temperature is required to yield void free parts.

It is desirable to cure AT resins completely to achieve optimum mechanical properties. Because of diffusion control of the cure reaction, the post-cure temperature cannot be substantially below the T_g of the fully cured system. The low T_g AT resins are usually based on flexible chemical structures which are not as thermally stable as the aromatic structures in the high T_g AT resins. Thus the post-cure temperature should be chosen as a compromise after considering the T_g of the fully cured product and the thermal stability of the resin.

The fully cured T_g's of ATP, ATB and ATS were determined to be 280°C, 275°C, and 345°C, respectively. The post-cure temperatures were chosen as 250°C for ATB and ATP, and 300°C for ATS.

The above discussion is intended to be as general as possible. It specifically addresses the processing conditions of continuous fibre reinforced composites. The choice of a cure cycle often requires

TABLE 4.2
PROCESSING CONDITIONS OF ATB USING 'NO-BAG' AUTOCLAVE TECHNIQUE

(a)	Hot melt prepreg the resin at 40°C.
(b)	Lay up prepreg tape at room temperature.
(c)	Apply 100 psi isotactic air pressure in autoclave.
(d)	Ramp temperature from 23°C to 170°C at 2°C/min.
(e)	Cool from 170°C to room temperature at 2°C/min.
(f)	Freestanding post-cure in air at 250°C.

consideration of the processing methodology and the specifics of the cured part. Table 4.2 shows an example of the processing conditions chosen for hot melt prepregging graphite fibre tows and fabrication of an ATB laminate.[30]

4.5. PROCESSING OF HIGH INITIAL T_g AT RESINS

4.5.1. Melt Temperature Range and Cure Temperature Range

It was pointed out in an earlier section that the DSC results of high T_g AT resins do not usually show appreciable exotherm until the experimental temperature is above the initial T_g. This, however, does not mean that the reaction will not occur at all at a temperature below the glass transition temperature. There will be a finite reaction rate below the T_g, the result of which is to advance the T_g of the material.

Since almost all processing requires the resin to be in the melt state initially, the initial T_g of the resin and the heating rate are both critical parameters to contend with in the processing of the high initial T_g AT resins. The processing windows of these resins are usually very narrow. The windows are loosely defined here as the period of time at the processing temperature in which the resin softens sufficiently for the required bulk flow and fusion to occur, before the cure causes an irreversible vitrification of the matrix.

In the case where the heating rate is too slow, such that the kinetic rate is always sufficiently advancing the T_g of the material to a value above the temperature of the work-piece, the material will never soften, making processing impossible.

For high initial T_g AT resins, the softening temperature range is often in the range where the kinetics of the acetylene reaction is very rapid. Because of the practical limitation of most conventional ovens and presses the heat-up rate achievable on the work-piece is limited. This makes the processing of some of the high T_g AT resins very difficult.

The feasibility of processing the high T_g AT resins into desired parts is often limited by the size of the parts and the capability of the processing equipment. For example, in attempts to process the neat resin, 'BATQ-H' (an AT resin with the quinoxaline backbone structure), it was found that the core of thick specimens may not be sufficiently fused although the surface appeared to have processed properly. Also, for the equipment used, it was found that inserting a cold mould charged with the uncured powder into a preheated press failed to process the resin properly. Instead, the mold had to be preheated to the processing temperature before being charged with the uncured resin.

While a lot of processing difficulties can be overcome by using equipment capable of transferring the thermal energy into the work-piece at faster rates, there exist properties intrinsic to a resin which potentially limit its processability. The initial T_g and the kinetics of the curing reaction are examples of these properties, and have been discussed. Thermal conductivity of the AT resins, which has not been studied thus far, can pose a practical limitation to the feasibility of processing some of the high initial T_g AT resins into large parts.

4.5.2. Viscosity

As with low T_g AT resins, viscosity profiles are important in the selection of proper cure conditions for the high initial T_g AT resins. Because of the overlap of the melting and curing temperature ranges, temperature scan experiments often fail to yield meaningful processing information. Instead, profiles from isothermal experiments at different cure temperatures are used to gain insight into their processing characteristics.

In our laboratory, the viscosity measurements were performed on the Rheometrics spectrometers (RMS-7200 and RDS) in the parallel plates mode. The environmental chamber and the equipment fixtures were preheated to the experimental temperature and allowed to stabilise for at least 15 min. The sample, in the form of a disc pressed from the powdery AT resin under high pressure at room temperature, was then quickly inserted between the plates. The resulting viscosity profile will show a rapid decrease at the beginning of the experiment because of the temperature rise of the sample. A minimum viscosity

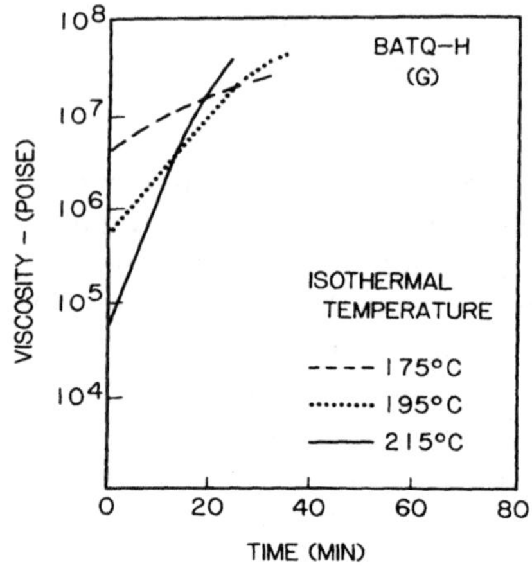

FIG. 4.4. Isothermal viscosity profiles of a quinoxaline based AT resin (BATQ-H).

level will be recorded at about 1 min after the sample insertion. (The exact time depends on the experimental temperature, and is reasonably reproducible.) Though the minimum viscosity recorded in these measurements is not strictly the initial viscosity of the resin at the experimental temperature, because of (i) uncertainty about when the sample has reached the experimental temperature, and (ii) the extent of the reaction taking place during the viscosity decreasing portion of the experiment, the value was treated as such for the cure cycle selection consideration. Figure 4.4 shows the isothermal viscosity profiles of the BATQ-H at different experimental temperatures.[28]

The data obtained from these experiments can be reconstructed to form an isoviscosity time–temperature chart, similar to the time temperature transformation (TTT) diagram proposed by Gillham.[29] The times needed to reach different levels of viscosity at various experimental temperatures were connected to yield a diagram similar to that shown in Fig. 4.5. The processing temperature would then be chosen based on the minimum viscosity and time required for the specific processing. This type of diagram is very useful in choosing the initial trial processing conditions for the high initial T_g AT resins. Since the information is obtained from experiments with unusually high heating rates, subsequent modifications of the trial conditions are not unusual. This is to accommodate the limitations of the processing equipment, and to achieve the desired quality in the processed parts.

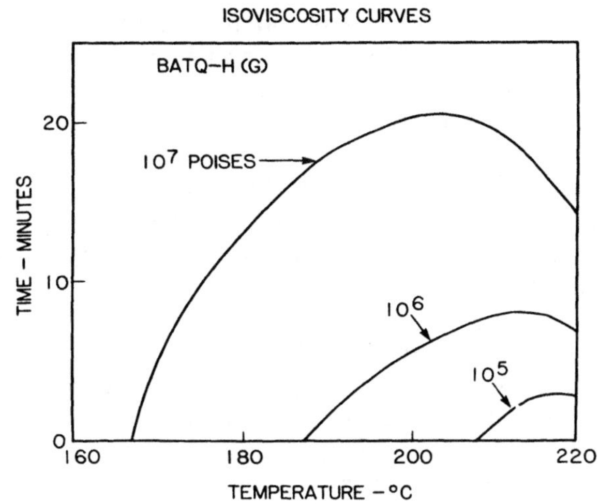

FIG. 4.5. Time–temperature plot of isoviscosity curves of a quinoxaline based AT resin (BATQ-H).

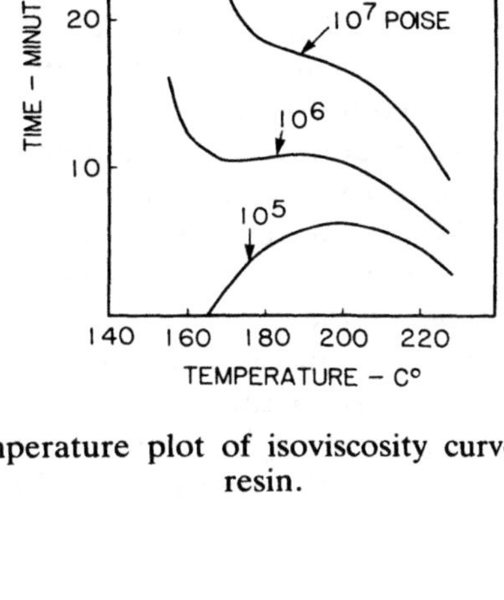

FIG. 4.6. Time–temperature plot of isoviscosity curves of an AT isoimide resin.

Figure 4.6 shows a similar diagram for another high initial T_g AT resin. The resin was an AT isoimide and had a T_g about 20°C lower than that of the BATQ-H. The high viscosity curve in this diagram shows a distinct upturn in the low temperature range. This upturn is as expected when the combination of high viscosity and low temperature significantly retarded the cure reaction rate. The reason why this feature is not in Fig. 4.5 is because the overlap of the melt temperature range and the reaction temperature range is so large that the viscosity upturn was not observed in the melt.

4.5.3. Means to Improve Processability

The primary reason that these high initial T_g resins are difficult to process is because of the overlap of their melt temperature range with the cure reaction temperature range. As was discussed in an earlier section, the reaction kinetics is relatively insensitive to the chemical structures of the chain between the acetylene end groups. Catalysts[31] and radical inhibitors,[32] however, have been used effectively to influence the kinetic rate of the acetylene reaction. Catalysts can lower the reaction on-set temperature, but this approach is of little use to improve the processability of the high initial T_g resins. Although radical inhibitors had been used to make the AT imide resin easier to process in the fabrication of adhesive lap shear joints, this approach has not been pursued far. If the concentration of the inhibitor is high, the thermal stability of the resulting matrix is compromised.

Most of the efforts to improve the processability of the high initial T_g AT resins attack the problem by lowering the initial T_g of the resins, whereby lowering the melt temperature range. Lowering the initial T_g of a resin will widen the processing window, even though the cure kinetics may not have been affected. For a given processing temperature, the lower initial T_g resin will have a lower initial viscosity. This will shorten the time required for the necessary bulk flow and consolidation steps, and simultaneously lengthen the time for the resin to reach the critical viscosity level, where further processing is no longer possible. In addition, a resin with a lower initial T_g can be processed at a lower temperature. This makes the kinetic rate slower and the processing window wider.

The aromatic structures of most of the high initial T_g resins were chosen primarily because of their thermal stability. Lowering the T_g of these resins by replacing the rigid aromatic structures with more flexible ones will often lower the thermal stability of the resins. The

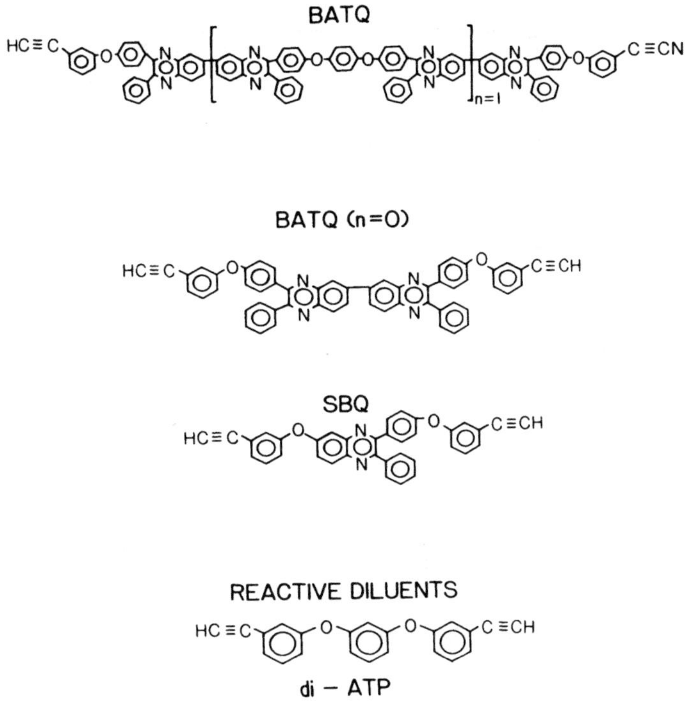

FIG. 4.7. Chemical structures of a series of quinoxaline based AT resins and a reactive diluent (ATP). The T_g's of the quinoxaline resins, BATQ-H, BATQ-H($n = 0$), and SBQ are 170°C, 110°C and 56°C, respectively.

alternative is to maintain the aromatic structure, but to reduce the chain length between the acetylene end groups.

This approach to lowering the initial T_g of the resins can be illustrated by the series of quinoxaline molecules synthesised in the Materials Laboratory of the US Air Force.[33] Figure 4.7 shows the chemical structures of this series of resins. The shortest structure in this series has an initial T_g of 56°C, which means that this resin can be processed just like the low initial T_g AT resins. The effectiveness of improving the processability of a resin by lowering the initial T_g can be illustrated by the viscosity curves shown in Fig. 4.8.

Lowering the initial T_g by shortening the chain length between the acetylene end groups is effective in improving the processability, but also tends to lower the toughness of the resulting matrices.[34]

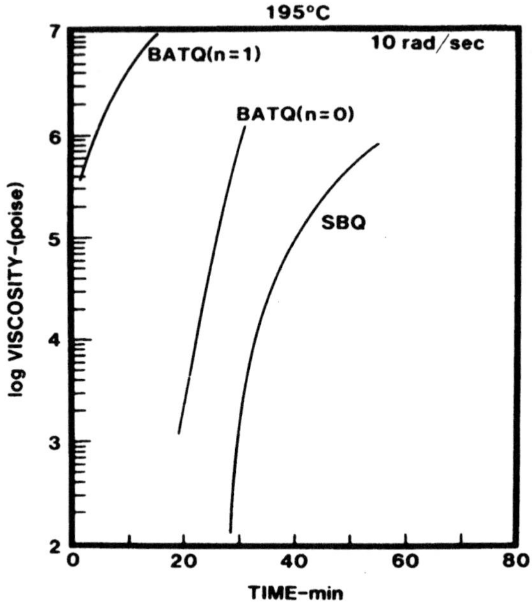

FIG. 4.8. Isothermal viscosity profiles of the quinoxaline based AT resins.

Another approach to lowering the initial T_g of a resin is the addition of reactive diluents.[35] The diluent can lower the initial T_g of the base resin by plasticisation. A proper choice of diluent will allow the plasticiser to co-react with the base resin during the processing, eliminating the tendency towards void formation found with conventional plasticisers and the lowering of the T_g of the cured matrix by the plasticisation effect of the residual diluent.

Acetylene terminated phenylene oxide (ATP),[15] the structure of which is shown in Fig. 4.7, has been used effectively for this purpose. The processability of BATQ-H can be tuned to suit specific processing requirements by controlling the concentration of the diluents. The effect of the diluent concentration on the viscosity profiles of the blends is illustrated in Fig. 4.9. By blending in a small amount of ATP satisfactory composite laminates can be made from BATQ-H with processing conditions that have failed with pure BATQ-H. Because of the acetylene end groups of ATP, these molecules are effectively incorporated into the cured network structure, with minimal effect on the final T_g and the thermal stability of the cured matrices.

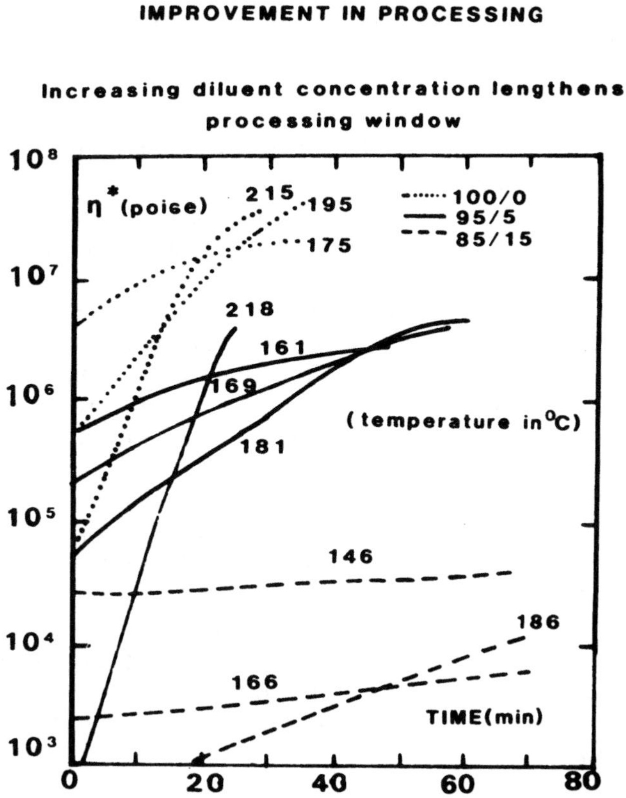

FIG. 4.9. Isothermal viscosity versus time profiles of BATQ-H with different concentrations of reactive diluent.

There is another approach that has been used successfully to improve the processibility of the AT imide resins. This method, however, is unique to the imide chemistry because of the possibility of synthesising a structure which is an isomer of the imide.[11] This isomer of imide, referred to as the isoimide, is more flexible, thereby lowering the initial T_g of the uncured resin and making the resin soluble in less exotic solvents (such as a 4 to 1 mixture of methyl ether ketone and toluene). The T_g lowering, and the improvement in solubility, greatly enhance the processability of the imide resin in terms of the prepregging operation and the subsequent consolidation steps. The isoimide reportedly will undergo thermally induced isomeric rearrangement during the processing, to convert back to the imide structure, thus theoretically it will retain all the desirable end-properties attributed to the original imide resin.

4.5.4. Processing Conditions

For most of the high initial T_g AT resins melt prepregging without severely limiting the subsequent processibility of the resulting prepreg tapes for continuous fibre reinforced advanced composite fabrication is extremely difficult. For some resins, such as BATQ-H, it is practically impossible.

The prepregging of these resins invariably requires solvents to achieve proper wetting of the fibres at a sufficiently low temperature. Some AT resins, e.g. the quinoxalines, can use common and easy-to-remove solvents, like methylene chloride; but some resins, such as the imides, can only be processed with exotic solvents, for example *N*-methylpyrrolidone (NMP).

Lamination of these prepregs invariably requires higher pressure, and faster heat-up rate, than can be provided by a conventional autoclave. Compression moulding in a heated hydraulic press is easier. While low viscosity melts are desirable to achieve good consolidation of the plies, they may not be ideal in terms of suppression of voids formed by any residual solvent left from the prepregging step. It was found that high pressure, at a temperature slightly below the initial T_g of the resin, can often achieve satisfactory consolidation and minimise void formation.[36]

Table 4.3 shows the prepregging and lamination conditions of a quinoxaline resin (BATQ-H) blended with 5% of reactive diluent.[30] The initial T_g of this blend is 156°C.

TABLE 4.3
PROCESSING CONDITIONS OF BATQ-H (BLENDED WITH 5% REACTIVE DILUENT)

(a)	Solvent prepreg the resin with dioxane as the solvent.
(b)	Degas prepreg 16 h at room temperature and 2 h at 80°C.
(c)	Layup prepreg in vacuum bag between metal caul plates.
(d)	Apply vacuum and 500 psi pressure.
(e)	Ramp to 150°C and hold 1 h.
(f)	Release vacuum.
(g)	Cure with 500 psi pressure for 1 h at 150°C, 170°C, 190°C, 210°C and 225°C.
(h)	Post-cure 4·5 h at 250°C under pressure in air.

4.6. REACTION MECHANISM

4.6.1. Initiation and Termination

Because of the intractable nature of the cured AT matrices the exact chemical structures of the acetylene reaction products are not known. Phenylacetylene was reported by Kern[37] to form polymeric complexes through thermally induced reactions. In 1974, it was proposed that acetylene reacted in bulk by trimerisation to form an aromatic ring structure.[7] A study using solid phase C13 NMR, with cross polarisation and magic angle spinning techniques, revealed that no more than 30% of the acetylene reacting in bulk trimerised to form aromatic structures.[38] It is likely that the rest of the acetylene groups form conjugated structures, (—C═C(R)—), as suggested by various studies.[39–41]

The thermally induced reaction was proposed to initiate through a biradical mechanism,[42] and to form the conjugated structure through a subsequent radical chain extension reaction.[43] The average degree of polymerisation by such chain extension reactions was found to be around six. The chain termination step was found to be a first order reaction, with a proposed intramolecular cyclisation mechanism. The free radical nature of the reaction is evidenced by the ESR study[44] which recorded a high concentration of free radicals in a cured AT matrix sealed in nitrogen. The kinetics of the AT reaction has been observed to be retarded in the presence of radical scavengers such as hydroquinone[32,45] or air.[46,47]

4.6.2. Proposed Network Structure

In spite of the fact that most of the acetylene groups react by 'chain extension' to form a conjugated structure, the molecular network resulting from the AT cure probably does not contain long segments of chains composed of these conjugated structures. When a molecular model of the proposed conjugated product of AT sulphone was built, it is obvious that the conjugated chain portion is functioning more like a crosslink than a chain because its occupied volume is substantially smaller than that of the rest of the molecule.[48] The laboratory model of the reaction product has the appearance of a multi-armed starfish (Fig. 4.10). The conjugated structure is in the centre, sterically shielded by the 'arms', each one of which, radiating from the centre, is the remaining portion of the molecular structure of the original AT resin. These starfish, or clusters, will have the number of arms

FIG. 4.10. A photograph of a molecular model of an ATS polyene with a degree of polymerisation of six, depicting the 'starfish' characteristics of the cluster.

corresponding to the degree of polymerisation of the conjugated polyene in the centres. Thus one can visualise that in the initial portion of the cure reaction, a number of clusters are formed in the resin bulk, with the average number of arms on these clusters being around six.

Although several spectroscopic techniques, e.g. NMR and FTIR[44,49–51] have been used to probe the chemical structures of the acetylenic reaction products of the AT resins, definitive determination is not yet in hand.

4.7. DEPENDENCE OF MECHANICAL PROPERTIES ON CURE

4.7.1. Extent of Cure

Most often, the selection of a cure cycle for a thermoset resin is limited to the consideration of the conditions at which the resin melt can be shaped into desirable forms, and simultaneously or subsequently, the time and temperature required irreversibly to vitrify the liquid resin into a glassy matrix. Post-cure of the cured matrix is often augmented to achieve improvement in mechanical properties, or to ensure that the glass transition temperature of the matrix is sufficiently advanced to satisfy the use temperature requirements. With the exception of systems like rubber modified epoxies, where the cure conditions can influence the development of the phase morphology, and thus in turn influence the mechanical performance of the cured matrices, the dependence of properties on the cure conditions of thermoset systems is not widely studied.

Some thermoset systems will show little dependence of the final mechanical properties on the cure conditions, but there is evidence that the properties of cured AT matrices, or at least of the AT sulphone, can be strongly influenced by the cure.[52]

First of all, the tensile properties' dependence on the extent of cure of the AT resins is quite different from that of the epoxy/amine system, tetraglycidylmethylene dianiline (TGMDA) with diaminodiphenylsulphone (DDS), which is the base of many resin systems currently being used in various aerospace applications. In a study of this epoxy/amine system,[45] it was found that the tensile strength and the elongation at break of an incompletely cured matrix were superior to those of a completely cured one, by as much as a factor of two. The tensile properties of AT sulphone, however, showed continuous improvement with increasing extent of cure.[53] The properties were rather poor in the intermediate cure state.

This difference is interesting from the material behaviour standpoint. It is also of practical importance in processing. Early attempts at fabricating advanced composites[54] with AT sulphone resins, using cure cycles similar to those used for epoxy systems, often resulted in poor quality laminates, with microcracks propagating through the cured matrices. Further investigation showed that the microcracks were formed before the post-cure. While an epoxy matrix at this stage of cure possesses excellent mechanical properties, an AT sulphone matrix at a comparable stage is too brittle to withstand the stresses induced by the volumetric shrinkage caused by the temperature decrease at the end of the cure cycle.

The difference between epoxy and AT sulphone resins in their tensile property dependency on the extent of cure can be rationalised through their network formation processes. Considering a simplified

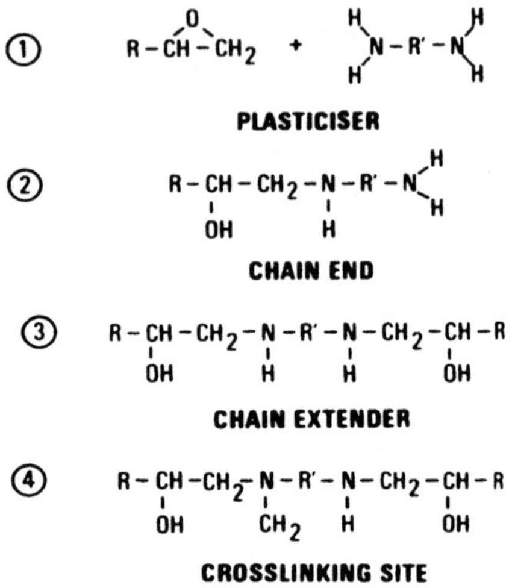

FIG. 4.11. Illustration of a simplified sequence of an epoxy/amine reaction. The role of the amine molecule changes from a plasticiser to a trifunctional, and finally, a tetrafunctional crosslink.

picture, without taking account of the many secondary reactions that are possible, the reaction sequence of a difunctional epoxy and a diamine can be outlined as shown in Fig. 4.11. After an amine molecule reacts with an epoxy molecule, the net effect is to increase the molecular weight of the epoxy group slightly. The matrix, if in the glassy state, will have no significant mechanical properties to speak of, because it is mostly held together through Van der Waals forces. The next likely reaction is that of the remaining primary amine with another epoxy group. The net effect of this reaction is a chain extension: different epoxy molecules have been joined by an amine molecule. Since the reaction of the primary amine will proceed more readily than the secondary amine, the early phase of the reaction will probably be generating a bulk consisting of many higher molecular weight linear chain molecules. When some of the secondary amine reacts, the role of the amine molecules will in effect change from being a chain extender to a crosslinking site. When a sufficient number of the amine molecules act as crosslinking sites, a continuous molecular network is formed. If the epoxy molecule is tetrafunctional like TGEBA, a continuous network can be formed, even before the secondary amine begins to react. The rationale for the improvement in properties with extent of cure in the beginning portion of the epoxy reaction is quite obvious.

As the reaction continues towards completion more and more of the amine molecules are acting as crosslinking sites. A tighter network with shorter molecular weights between crosslinks is formed, resulting in a deterioration in mechanical properties.

AT resins, however, form multi-armed clusters in the initial stage of the cure, as was discussed in an earlier section. Although the remaining acetylene groups on the tips of the arms of these clusters are still unreacted, and are capable of further reaction, it is more likely the unreacted monomers will react first because of their greater mobility. Thus the resulting matrix in the initial stage of cure consists of many disconnected clusters, which contribute little to the mechanical integrity of the bulk (Fig. 4.12). At a later stage of cure, after sufficient reaction between the clusters has occurred, a continuous network will eventually be formed.

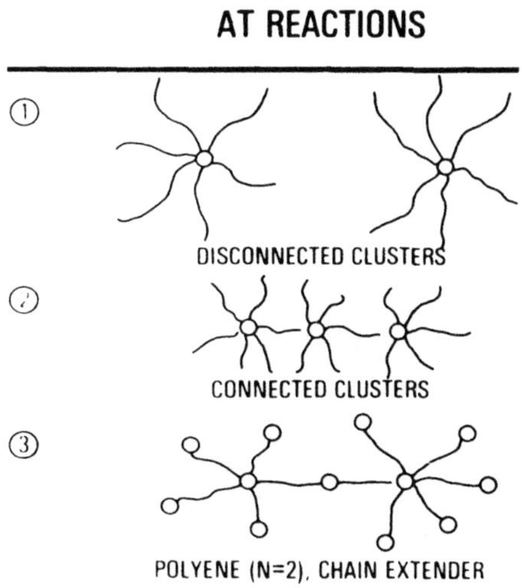

FIG. 4.12. Illustration of the proposed AT network formation process.

It should be pointed out that while most of the polyenes act as multi-functional crosslinking sites, with their functionality the same as the degree of polymerisation of the polyenes, those polyenes that have a degree of polymerisation of two are in effect chain-extenders.

4.7.2. Cure Environment

Because the acetylene reaction is radical initiated it is not surprising that the cure kinetic rate is retarded when the resin was cured in air. In a study comparing the tensile properties of cured AT sulphone specimens cured in air with those cured in nitrogen,[47] it was found that the tensile strength and elongation at break of those cured in air were consistently higher (Fig. 4.13). The exact cause of this difference has yet to be determined. One cannot exclude the possibility that the presence of air has drastically changed the composition of the reaction products.

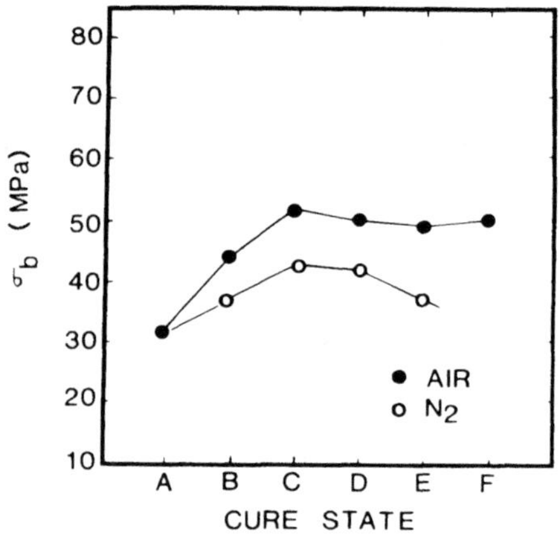

FIG. 4.13. Tensile strength of ATS specimens cured with identical cure cycles, except under different environments (air and nitrogen). The x-axis displays the cure cycles in increasing extent of cure.

Another plausible explanation is that the oxygen in air has interfered with the chain propagation step, by annihilating the free radicals on the propagating chains. This annihilation, in addition to reducing the kinetic rate, will shift the distribution curve of the degree of polymerisation of the polyenes towards lower numbers, and in effect, will decrease the functionality of the crosslinks, resulting in a less strained network. Moreover, if the number of polyenes with a degree of polymerisation of two increases the effective molecular weight between crosslinks in the resulting network will increase also. All these factors could contribute to the improvement in tensile properties.

4.7.3. Low Temperature Cure

Table 4.4 shows the tensile properties of two sets of ATS specimens, post-cured under identical conditions, but with one set being cured at a lower temperature for an extended period of time (cure cycle B) and the other only for a short period of time at a higher temperature (cure cycle A). The high temperature properties are very similar, but the room temperature data clearly show that the low temperature, long time cure has improved the tensile properties of the cured matrix.

As was mentioned in an earlier section, previous attempts to fabricate ATS graphite-reinforced advanced composite laminates resulted in microcracks in the matrices. The tensile properties of the neat resins, cured with cycle B, are nearly a factor of two better than those of those cured with the earlier cure cycle. Using cure cycle B, cross ply (0°/90°) laminates of ATS with continuous graphite-fibre reinforcement have been successfully fabricated without evidence of microcracks.[55]

TABLE 4.4
TENSILE PROPERTIES OF ATS FROM DIFFERENT CURE CYCLES

Cycle	*Conditions*	*Comment*	*Tensile strength*	*Strain at break*
A	120°C/1 h	Degassing	7·8 ksi	1·9%
	150°C/3/4 h	Cure in mould	(54 mPa)	
	170°C/3/4 h	Post-cure		
	300°C/10 h	Post-cure		
B	120°C/1 h	Degassing	9·8 ksi	2·3%
	140°C/24 h	Cure in mould	(68 mPa)	
	200°C/1 h	Post-cure		
	300°C/10 h	Post-cure		

Again the reason for the improvement of the tensile properties could have been the change in the chemical reactions of the acetylene at different reaction temperatures. However, spectroscopic studies of the acetylene reaction has, thus far, yielded no evidence to support such an explanation.[56]

Another plausible explanation is that the improvement is due to the vitrification process in cure cycle B being under a lower kinetic rate condition. The vitrification process is very similar to the process whereby a low molecular weight polymer melt is transformed into the glassy state by lowering the temperature of the polymer across the glass transition temperature region. If the rate of the temperature decrease is too fast, or in the extreme case if the sample is in effect quenched, the internal stresses due to the rapid volume shrinkage can cause the matrix to be very brittle. When the vitrification process is slow, there is more time for the network to relax the stresses induced by the cure reaction, and by the volume shrinkage caused by the transformation of the melt to a glass.

Still another explanation can be forwarded, using the argument that a higher percentage of the reaction in cure cycle B occurs under a higher viscosity condition. In cure cycle B, the initial viscosity of the resin is higher and the viscosity build-up occurs at a lower extent of cure. Under the high viscosity environment, molecular motions are more restricted, and in effect, retard the chain propagation step of the reaction mechanism. As was discussed earlier, interference of the chain propagation can conceivably shift the distribution of the degree of polymerisation of the polyenes to lower numbers, leading possibly to a less stressed network, and to better tensile properties.

The above speculations, however, do not satisfactorily account for the lack of difference between the high temperature tensile properties resulting from the two cure cycles.

It is not clear at this point whether the observed dependency of the room temperature tensile properties on cure conditions is unique to the sulphone resins, or is generally true for all AT resins. Slight improvement was observed when AT bis-phenol-A resins were cured with a long, low temperature cure cycle.[57,58] Using the same long, slow low temperature approach, the tensile properties of ATP[28] showed an improvement by a factor of two over the reported value of an earlier study.[59]

It is interesting to point out that the specimens cured with cure cycle A and B, listed in Table 4.4, in addition to showing differences in room temperature tensile properties, exhibit drastically different property retention characteristics under high temperature long-term ageing conditions.[52] Figure 4.14 shows the tensile strength of the specimens cured with the two different cure cycles, as a function of ageing time at 450°F (232°C) in air. The specimens cured with cure cycle B (long time at low temperature) showed property deterioration after about 4 weeks of exposure, but cure cycle A gave good retention characteristics even after 8 weeks.

The stability of the polyene crosslink is largely due to the delocal-

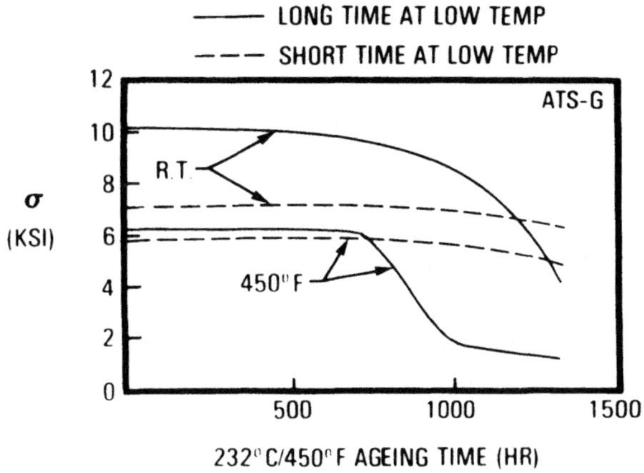

FIG. 4.14. Tensile strength retention of ATS as a function of ageing time at 232°C (450°F) in air.

ised nature of the electrons in the conjugated chain. The stability will decrease with decreasing length of the polyene. Although it is still speculation, the data shown in Fig. 4.14 appears to be consistent with the argument that cure cycle B causes a shift of the distribution curve of the degree of polymerisation of the polyene to a lower value.

4.8. CONCLUSIONS

Even though the thermal characteristics of the curing reaction are relatively insensitive to the chemical structures of the AT resins, the curing characteristics of these resins can be quite dependent on the initial T_g of the resins. For the low initial T_g AT systems, the resin melt temperature range is sufficiently separated from the cure reaction temperature range that the processing of these resins are quite simple. But for the high initial T_g AT systems, the overlap of the resin melt temperature range and the cure reaction range can make the processing of these resins difficult. Improvements in processing can be achieved by decreasing the overlap of the two temperature ranges, and methods for lowering the melt temperature range of these AT resins, thus widening the processing windows, have been discussed. The mechanical properties of AT resins have a different dependency on the extent of cure from that of an epoxy/amine system. The difference can be explained by the different processes by which the molecular networks are formed during cure. Dependency of mechanical properties and stability on cure conditions have also been observed in an AT system, and the possible causes of this behaviour were also discussed.

ACKNOWLEDGEMENTS

The author would like to thank Dr I. J. Goldfarb, Dr F. E. Arnold, Dr T. E. Helminiak and Dr C. C. Kuo for the many helpful discussions about this technology area throughout the years. These discussions have contributed to a great extent to the content of this chapter.

REFERENCES

1. Lee, C. Y-C., Goldfarb, I. J., Arnold, F. E. and Helminiak, T. E., *ACS Org. Coat. Appl. Polym. Sci. Prepr.*, **48,** 903 (1983).
2. Hergenrother, P. M., in *Encyclopedia of Polymer Science and Engineering,* John Wiley Publishing Co., Chichester, in press.
3. Lee, C. Y-C., Goldfarb, I. J., Helminiak, T. E. and Arnold, F. E., *Proc. 28th Nat. SAMPE Symposium,* p. 699, April, 1983.
4. Lau, K. S. Y., Kelleghan, W. J., Boschan, R. H. and Bilow, N., *J. Polym. Sci. Polym. Chem. Ed.* (in press).
5. Jabloner, H. and Cessna, L. C., *J. Elastomers Plast.*, **6,** 103 (1974).
6. Bilow, N., Landis, A. L. and Miller, L. J., US Patent 3,845,018 (1974).
7. Landis, A. L., Bilow, N., Boschan, R. H., Lawrence, R. E. and Aponyi, T. J., *Polym. Prepr.,* **15** (2), 533 and 537 (1974).
8. Bilow, N. and Landis, A. L., *Polym. Prepr.*, **19** (2), 23 (1978).
9. Bilow, N., Keller, L. B., Landis, A. L., Boschan, R. H. and Castillo, A. A., *Sci. Adv. Mat. Proc. Eng. Ser.*, **23,** 791 (1978).
10. Hanky, A. O. and St Clair, T. L., *Sci. Adv. Mat. Proc. Eng. Ser.*, **28,** 711 (1983).
11. Landis, A. L. and Naselow, A. B., *Nat. SAMPE Tech. Conf. Ser.*, **14,** 236 (1982).
12. Hedberg, F. L. and Arnold, F. E., *Polym. Prepr.,* **18** (1), 826 (1977).
13. Hergenrother, P. M., *Org. Coat. Plast. Chem. Prepr.,* **36** (2), 264 (1976).
14. Kovar, R. F., Ehlers, G. F. L. and Arnold, F. E., *Polym. Prepr.,* **16** (2), 246 (1976).
15. Maximovich, M. G., Lockerby, S. C. and Arnold, F. E., US Patent 4,131,625 (1978).

16. Loughran, G. A. and Arnold, F. E., *Polym. Prepr.*, **21** (1), 199 (1980).
17. Hedberg, F. L., Unroe, M. R., Lindley, P. M., Hunsaker, M. E. and Feld, W. A., *ACS Polym. Prepr.*, **24** (1), 153 (1983).
18. Loughran, G. A., Reinhart, B. A., Arnold, F. E. and Soloski, E. J., *ACS Org. Coat. Plast. Chem. Prepr.*, **43,** 777 (1980).
19. Evers, R. C., Moore, G. J. and Burkett, J. L., *ACS Polym. Prepr.*, **22** (1), 7 (1981).
20. Roller, M. B., *Polym. Eng. Sci.*, **15,** 406 (1975).
21. Gillham, J. K., *Soc. Plastics Eng. Techn. Pap.*, **22,** 239 (1976).
22. Lee, C. Y-C. and Goldfarb, I. J., *Polym. Eng. Sci.*, **21,** 390 (1981).
23. Tung, C. Y. and Dynes, P. J., *J. Appl. Polym. Sci.*, **27,** 569 (1982).
24. Flynn, J. H. and Wall, L. A., *J. Res. Nat. Bur. Stds,* **70A** (6), 487 (1966).
25. Goldfarb, I. J. and Adams, W. W., *ACS Org. Coat. Plast. Chem. Prepr.*, **45,** 133 (1981).
26. Goldfarb, I. J., AFWAL/MLBP, private communications.
27. Denny, L. R. and Lee, C. Y-C., AFWAL-TR-84, in preparation.
28. Kuo, C. C. and Lee, C. Y-C., AFWAL-TR-84, in preparation.
29. Gillham, J. K., *Polym. Eng. Sci.*, **19,** 676 (1979).
30. Dynes, P. J., Rockwell Science Center, Thousand Oaks, California, Interim Technical Report No. 4 and No. 7, US Air Force Contract F33615-82-C-5064 (1983).
31. Lucarelli, M. A., Jones, W. B., Jr, Picklesimer, L. G. and Helminiak, T. E., *ACS Polym. Prepr.*, **22** (1), 19 (1981).
32. Kuhbander, R. J. and Aponyi, T. J., *SAMPE Nat. Meeting Proc.*, **11,** 295 (1979).
33. Hedberg, F. L. and Arnold, F. E., AFML-TR-78-142 (1982).
34. Jones, W. B., Jr and Helminiak, T. E., AFWAL/MLBP, private communication.
35. Kuo, C. C. and Lee, C. Y-C., *ACS Polym. Prepr.*, **24** (2), 7 (1983).
36. Sikka, S. S. and Goldfarb, I. J., Air Force Technical Report, AFWAL-TR-81-4101 (1981).
37. Kern, R. J., *J. Polym. Sci.*, A-1, **7,** 621 (1969).
38. Sefcik, M. D., Stejskal, E. O., McKay, R. A. and Shaefer, J., *Macromolecules,* **12** (3), 423 (1979).
39. Simionescu, C. I., Percec, V. and Dumitrescu, S., *J. Polym. Sci. Polm. Chem. Ed.*, **15,** 2497 (1977).
40. Simionescu, C. I. and Percec, V., *J. Polym. Sci. Polym. Sym.*, **67,** 43 (1980).
41. Ratto, J. J., Dynes, P. J. and Hamermesh, C. L., *J. Polym. Sci. Polym. Chem. Ed.*, **18,** 1035 (1980).
42. Pickard, J. M., Jones, E. G. and Goldfarb, I. J., *Macromolecules,* **12** (5), 895 (1979).
43. Amdur, S., Cheng, A. T. Y., Wong, C. J., Ehrlich, P. and Allendoerfer, R. D., *J. Polym. Sci. Polym. Chem. Ed.*, **16,** 407 (1978).
44. Sandreczki, T. C. and Lee, C. Y-C., *ACS Polym. Prepr.*, **23** (2), 185 (1982).
45. Lee, C. Y-C., AFWAL/MLBP, unpublished data.
46. Lee, C. Y-C., *J. Appl. Polym. Sci.*, **27,** 407 (1982).

47. Kuo, C. C. and Lee, C. Y-C., *ACS Org. Coat. Plast. Chem. Prepr.*, **47,** 144 (1982).
48. Helminiak, T. E., AFWAL/MLBP, private communications.
49. Lind, A. C. and Lee, C. Y-C., *ACS Polym. Prepr.*, **23** (2), 185 (1982).
50. Levy, R. L. and Lee, C. Y-C., *ACS Polym. Prepr.*, **23** (2), 181 (1982).
51. Koenig, J. L., Case Western Reserve University, Air Force Technical Report, AFWAL-TR-83-4064 (1983).
52. Lee, C. Y-C., *ACS Polym. Mat. Sci. Eng. Proc.*, **51,** 405 (1984).
53. Lee, C. Y-C., Kuo, C. C. and Lee, N. C., Air Force Technical Report, AFWAL-TR-82-4098 (1982).
54. Hartness, T., University of Dayton Research Institute, private communications.
55. Leung, C., Rockwell Science Center, Thousand Oaks, California, Interim Report of Air Force Contract No. F33615-80-C-5142.
56. Lind, A. C., Levy, R. L. and Sandreczki, T. C., McDonnel Douglas Research Lab. St Louis, Missouri, Air Force Contract No. F33615-80-C-5170.
57. Abrams, F. L. and Browning, C. E., *ACS Org. Coat. Plast. Prepr.*, **48,** 908 (1983).
58. Lee, C. Y-C., Denny, L. R. and Goldfarb, I. J., *Polym. Prepr.*, **24** (1), 139 (1983).
59. Cotts, D., SRI International, Menlo Park, California, Air Force Technical Report, AFWAL-TR-81-4146 (1981).

Chapter 5

EFFECT OF CHEMORHEOLOGY ON EPOXY RESIN PROPERTIES

A. APICELLA

Polymer Engineering Laboratory, Department of Material and Production Engineering, University of Naples, Italy

SUMMARY

Chemoviscosity is the term which denotes the variations in viscosity induced by chemical reactions, and chemorheology is the science which studies the viscoelastic behaviour of reacting systems. This chapter briefly reviews some of the theoretical and experimental aspects of the chemorheology of epoxy resin systems.

5.1. INTRODUCTION

The synthesis and the processing of thermoplastics are two distinct operations. These polymeric materials soften and flow on heating but return to the solid state when cooled. Thermosets, on the other hand, are polymerised and processed in a single irreversible operation which transforms a low molecular weight liquid into a cross-linked polymer. Although thermosets, as solid materials, should potentially have superior engineering properties they often do not possess good reproducibility in their characteristics, in contrast to thermoplastic polymers, because of the less controlled and more complex formation process. The cure of a reactive prepolymer involves the transformation of low molecular weight monomers or oligomers from liquid to rubber and solid state, as a result of the formation of a polymeric network by the chemical reaction of the reactive groups in the system. Gelation and vitrification, which are two macroscopic phenomena encountered during this process, strongly alter the viscoelastic behaviour of the material.

Gelation, associated with a dramatic increase in viscosity, occurs at a degree of reaction calculable for each reacting system according to the theory first proposed by Flory.[1] On a molecular level, the increase in viscosity corresponds to the increase in molecular weight, and to the incipient formation of infinite branched molecules. The growth and branching of the polymer chains occur in the liquid state, while the system is still soluble and fusible, whereas the infinite network is developed after the gel point, by intramolecular reactions of the branched molecules, finally leading to an insoluble cross-linked solid.

Vitrification is the formation of a glassy solid, which usually follows gelation, occurring as a consequence of the network becoming denser through further cross-linking. Vitrification may prevent further reaction by reducing the mobility of the unreacted functional groups. Although chemical control is the basic assumption of all statistical treatments of curing, in some cases, cure may also be controlled by physical factors, such as diffusion constraints in the glass transition region, leading to the formation of nonhomogeneous structures, which indirectly alter the physical properties and durability of cured epoxy resins.[2] Knowledge of the structural parameters of the cured systems, as a function of the reactivity of the functional groups and their stoichiometric ratios, is important for the elucidation of the curing reaction, the control of processing, and application properties.

The chemical characterisation and reproducibility of thermosetting epoxies, which are the dominant form of polymeric matrices for carbon fibre composites, is therefore an important area of concern.

Completed studies of graphite epoxy composites for aerospace applications have revealed two dominant classes of hygrothermal degradation mechanisms: microscopic defects in the molecular network structure of the matrix phase, and macroscopic defects such as voids, bubbles, debonded and broken fibres produced during the processing of the composite.

The influence of the matrix nodular structure on the durability in aggressive environments, the mode of failure and mechanical properties of TGDDM–DDS* based composites has been reported in the

* Tetraglycidyl diamino diphenyl methane cured with diphenyl dimethyl sulphone.

literature by Morgan *et al.*[3] and Schneider *et al.*[4] It then becomes difficult to establish a close relationship between the chemical composition, the processing variables and the optimum characteristics in real life without accounting for the influence of the time–temperature path on the molecular structure during the early stages of cure, when gelation occurs, and in the final postcuring of these materials.

5.2. MORPHOLOGICAL AND CHEMICAL ASPECTS OF THE CURE OF THERMOSETS

5.2.1. Early Theories

Statistical theories have been used to describe the curing of epoxy resins. The results, however, were somewhat invalidated by the over-simplified reaction schemes adopted. The importance of deviations from the ideal network structure in cured epoxies, due to stoichiometric imbalance or to incomplete reaction, is recognised in the literature.[5] The development of modern statistical theories of branching processes, employing cascade substitution and probability generating functions, deals with the complex problem of network formation.[6–9] These models, derived from the Flory–Stockmayer tree-like pre-gel polymerisation, with random circuit closing, take into account the different reactivities of the active groups, the reaction mechanisms and a treatment of the conformational cyclisation, offering considerable structural information on the pre-gel and post-gel states. The principal concepts used in the development of such models will be briefly discussed.

Although polymers possess complex spatial distributions, a combination of a limited number of different elementary units connected by covalent bonds suffices for the identification of the entire polymer molecule. The simplest conceivable pattern, where the units are connected in linear sequence, describes linear polymers. Alternatively, non-linear or branched structures are obtained from systems containing some structural units of valency greater than two. Highly complicated molecules may further interconnect to yield a space network.

The structural units have a direct correspondence with the monomeric compounds employed in the preparation of the polymer. The interlinking capacity of the monomers is described by the concept of functionality first introduced by Carothers.[10,11] Monomers which are linked with only two other monomers after the polymerisation are termed bifunctional, while those joined to more than two monomers are classified as polyfunctional. Obviously, linear polymers are obtained from bifunctional monomers, while the synthesis of a non-linear polymer needs the presence of some monomers of functionality greater than two.

A useful differentiation between condensation and addition polymers has been proposed by Carothers.[10,11] In condensation polymers the molecular formula of the structural units differs from that of the monomers used for polymerisation but in addition polymers the molecular formulae of the structural units and of the monomers are the same.

However, it is not significant whether or not the structural units differ from the monomer. The differentiation between condensation and addition polymers lies in the marked differences between the mechanisms of polymerisation. Condensation proceeds by stepwise intermolecular reaction of the functional groups, while addition polymerisation is characterised by a chain mechanism, involving an active centre. In the latter, the addition of other monomers follows in rapid succession the primary activation of a monomer, until the active sites of the growing chain are deactivated. The reacting mixture is hence essentially composed of unreacted monomers and high polymers. The synthesis of a condensation polymer, conversely, is accomplished by a series of independent reactions involving monomers, dimers, trimers, etc., and, in contrast to addition polymerisation, where high molecular weight molecules are present from the beginning, completion of the reaction should be achieved in order to obtain high average molecular weight.

The mechanism of addition of ring compounds such as epoxides, however, does not resemble addition polymerisation, as previously defined. This polymerisation usually proceeds by an interchange reaction, induced either catalytically or by end-group producing substances. Ring to chain polymerisation proceeds kinetically according to a stepwise mechanism of intermediate character between condensation and addition because of the similar nature of the initial step and subsequent monomer addition.

In the case of non-linear condensation the growth of the macromolecule, which is not restricted to two directions, increases its functionality continuing the building of the structure to yield finally on a molecular level, an essentially infinite network. A rather sudden transformation from a viscous liquid to an infinitely viscous gel characterises the incipient formation of the spatial network. At this point only a small fraction of the material is insoluble and infusible while the remainder is still soluble. Two portions of the polymer can be then observed: the soluble 'sol' and the insoluble 'gel'. As the polymerisation proceeds the portion of gel increases at the expense of the sol. The theory presented by Flory[1] provides the basis for a description of three-dimensional polymerisation by polyfunctional reactions, together with a quantitative relationship governing the location of the gel point, the transformation of sol to gel, and the molecular distribution.

5.2.2. Cure Reactions

An outline knowledge of the mechanisms of reaction leading to the crosslinking of epoxy prepolymers may be useful to enable appropriate considerations to be used in the theoretical modelling of network formation. Chain extension and crosslinking of epoxide resins depend upon the reaction of epoxy groups with themselves and with the hydrogen of donor compounds. The reaction may be promoted by heat alone, by catalysts, which are not part of the resin molecule, or by agents which chemically interact with the epoxides. Figure 5.1 shows some of the most commonly used epoxy resins and hardeners.

Amines are usually used as curing agents for epoxies; tertiary amines catalytically promote epoxide homopolymerisation, while primary and secondary amines predominantly act by addition to the epoxide groups. The nucleophilic addition of primary amines to the epoxide results in the formation of amino alcohols. When primary diamines are used further chain extension first occurs followed by branching and crosslinking. The tertiary amine may induce crosslinking by forming an intermediate polyradical, which activates the polymerisation of the epoxy terminated resins and finally leads to a crosslinked mass. In this case amines function as catalysts for the etherification of the epoxides. Phenolic tertiary amines, the most commonly used catalysts of this type, may also be incorporated in the structure of some epoxy resins, such as tetraglycidyl diamino diphenyl methane (TGDDM). The hydroxyls formed during reaction with amines possess a further catalytic effect,[12] due to the hydrogen bonds formed on the epoxy oxygen, which may promote the action of the amines and lead to autocatalytic behaviour. Aliphatic amines react

Resins:

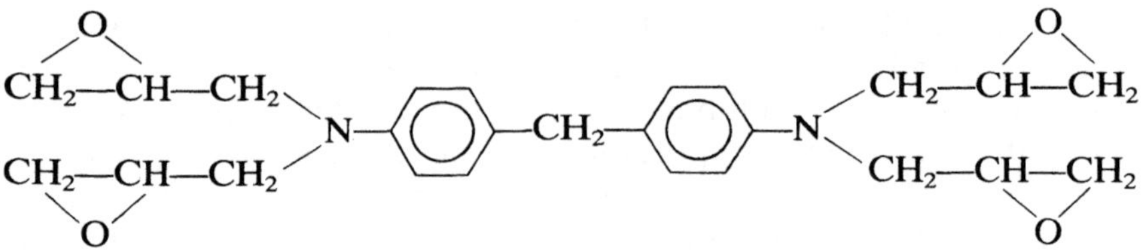

(TGDDM)

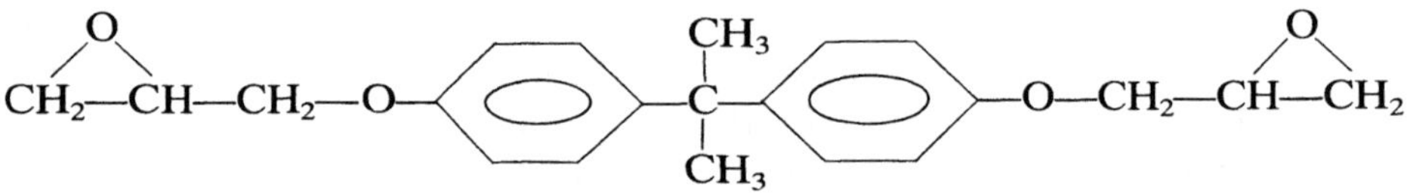

(DGEBA)

Hardeners:

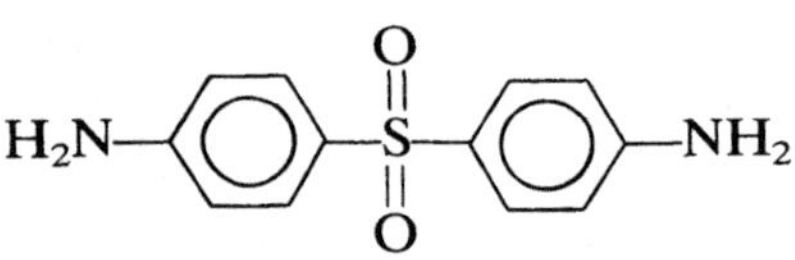

(DDS)

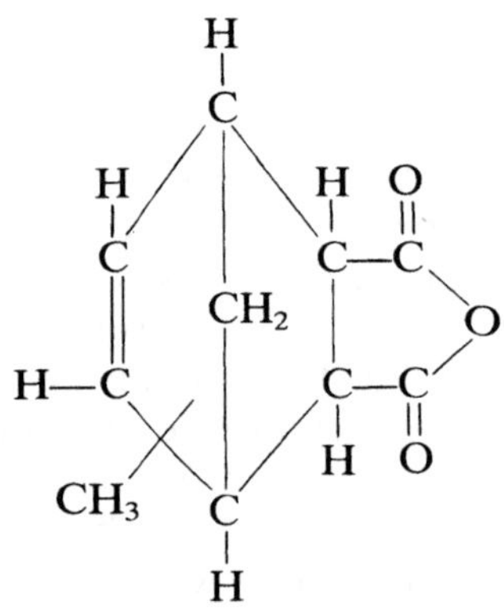

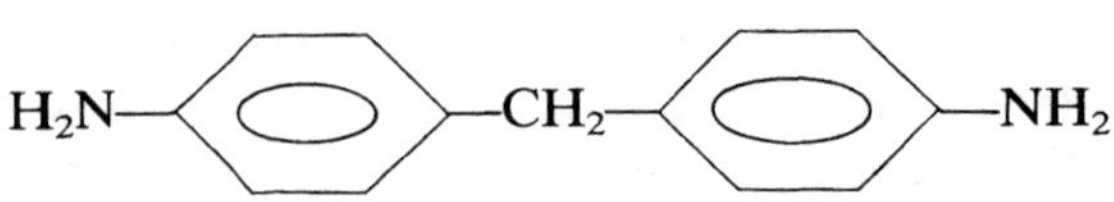

(DDM)

(MNA)

FIG. 5.1. Some of the most commonly used epoxy resins and hardeners. Tetraglycidyl diamino diphenyl methane (TGDDM), diglycidyl ether of bisphenol A (DGEBA), diamino diphenyl sulphone (DDS), diamino diphenyl methane (DDM) and methylated maleic acid adduct of phthalic anhydride (methyl nadic anhydride, MNA).

more rapidly than aromatic amines, and are therefore preferred for adhesive formulations, while the latter are often used in the matrices of epoxy prepregs because of their longer pot lives.

A second class of hardeners for epoxides are the organic acids or, more commonly, their anhydrides. These are able to induce chain extension and crosslinking, both by catalysing the homopolymerisation and, by reacting, to form polyesters, with the epoxy groups or hydroxyls derived therefrom. The fact that the optimum properties of the resin are often achieved using less than stoichiometric quantities confirms that catalysis is an important part of the reaction. Since it is the organic acid which is directly involved in the reaction with the epoxide, several reactions have to occur during the anhydride cure of an epoxy resin, namely:

(a) the formation of a monoester containing a free carboxyl group, resulting from the reaction of the anhydride with an hydroxyl (initially present as an impurity, or introduced with promoting agents)
(b) the main reaction of the carboxyl, formed in the previous step, with the epoxide to form a diester and a new hydroxyl (this reaction therefore increases the hydroxyl concentration)
(c) the polyetherification of epoxides induced by the reaction with the hydroxyls (since the ring opening by hydroxyl results in the formation of a new hydroxyl, this reaction does not alter their concentration in the system)

Cure with anhydrides needs, therefore, the use of compounds containing primary alcohols in order to activate the crosslinking reactions. Furthermore, the rate of cure with anhydrides may be significantly increased by agents such as boron trifluoride-amine complexes, organic sulphides and tertiary amines. Such compounds, however, induce catalytic homopolymerisation as in the case of the amines.

5.2.3. Theoretical and Experimental Determination of the Gelation Limits

Successful processing of thermosetting resins requires an understanding of the relationship between rheological behaviour, curing kinetics and changes in the molecular weight distribution. Consider the copolymerisation of a mixture of 100 parts of tetraglycidyl diamino diphenyl methane (TGDDM) and 35 parts of diamino diphenyl sulphone (DDS) as an example to define the conditions under which an infinite network is generated. Assume that all functional groups actually involved in the reaction possess equal reactivity (independently of the degree of reaction) and that condensation preceding gelation occurs only between different macromolecules (i.e. there is no cyclisation). The epoxy functional groups may be capable of further reaction with other epoxides, if combined with the hydroxyl formed in the ring opening, and the diamine possesses four reactive hydrogens. The TGDDM and the DDS molecules, however, at least in the initial stages of the polymerisation, should be considered tetrafunctional and difunctional, respectively, due to the strong differences in reactivity between the epoxide and primary amine groups and the hydroxyls and secondary amines.[13–15] Nevertheless, the gelation theory is applicable in principle to any polyfunctional system if the magnitude of such effects can be established by appropriate information on the single reaction kinetics.

In order to determine the conditions for formation of an infinitely large branched structure Flory[1] introduced a convenient quantity, the branching coefficient α, defined as the probability that a given functional group of a branched unit of functionality greater than two leads, via a chain of bifunctional units, to another branched molecule. In our case, if we denote by P_a and P_e, respectively, the fractions of primary amine and epoxy groups which have reacted, the branching coefficient is

$$\alpha = P_a P_e = rP_a = P_e/r \tag{1}$$

where r is the ratio of amine (m_a) to epoxy (m_e) groups initially present in the reactive mixture

$$r = m_a/m_e = gM_a/fM_e \tag{2}$$

with f and g the functionalities of the M_e and M_a moles of TGDDM and DDS molecules.

A convenient critical condition for the incipient formation of an infinite network is given in terms of the branching coefficient by

$$\alpha_c = 1/(f-1) \tag{3}$$

and therefore as a function of the mixture composition and epoxy conversion

$$P_{ec} = \sqrt{r/(f-1)} = \sqrt{M_a g/(M_e f(f-1))} \tag{4}$$

In particular, for the system presented here[16] $\alpha = P_e^2/0{\cdot}320$, $\alpha_c = 0{\cdot}33$ and $P_{ec} = 0{\cdot}315$.

Generally, the molecular constitution of the products of condensation of f-functional with bifunctional units is described by the complexity distribution and by the average number of bifunctional units per chain. As the reaction progresses larger species are formed at the expense of smaller ones. The theoretical concentration of monomers, which is always greater than all the other species, decreases continuously, leading to the formation of dimers, trimers, etc., whose concentration reaches a maximum and then decreases. The curves of the weight fraction of various species continue through the gel point without discontinuity. Flory[17] first derived the size distribution relationships for tri- and tetra-functional branching units that were subsequently generalised by Stockmayer[18] for f-functional units. The weight distribution function of x-mers in a poly-f-condensation is

$$W_x = [(1-\alpha)/\alpha]((fx-x)!\,f)/((x-1)!\,(fx-2x+2)!)\beta^x \qquad (5)$$

where

$$\beta = \alpha(1-\alpha)^{f-2}$$

The epoxy conversion, and hence the branching coefficient present in the above expressions, is usually derived from experimental techniques such as infrared spectroscopy, thermal analysis or liquid chromatography. Differential scanning calorimetry (DSC) is often used for the indirect determination of the advancement of the cure of thermosetting systems, with the assumption that the heat evolved is proportional to the extent of reaction. Several kinetic methods of determination of the rate constants and activation energies of the cure reactions are based on the same assumptions.[19,20] These assumptions, however, often do not hold for epoxy systems. These methods, in fact, assume simple nth order kinetics, with the temperature-dependent rate constant given by an Arrhenius expression, although multiple reactions and autocatalytic behaviour often characterise epoxy reactions.[16,21] DSC analyses, however, were successfully applied to complex systems when the kinetics of the single reactions were independently determined.[15,16] DSC traces, as obtained in complex cure cycles, e.g. the upper part of Fig. 5.2, may therefore be used to determine the actual epoxy conversion kinetics, lower part of Fig. 5.2, thus leading to a theoretical determination of the molecular composition, according to eqns (1)–(5).

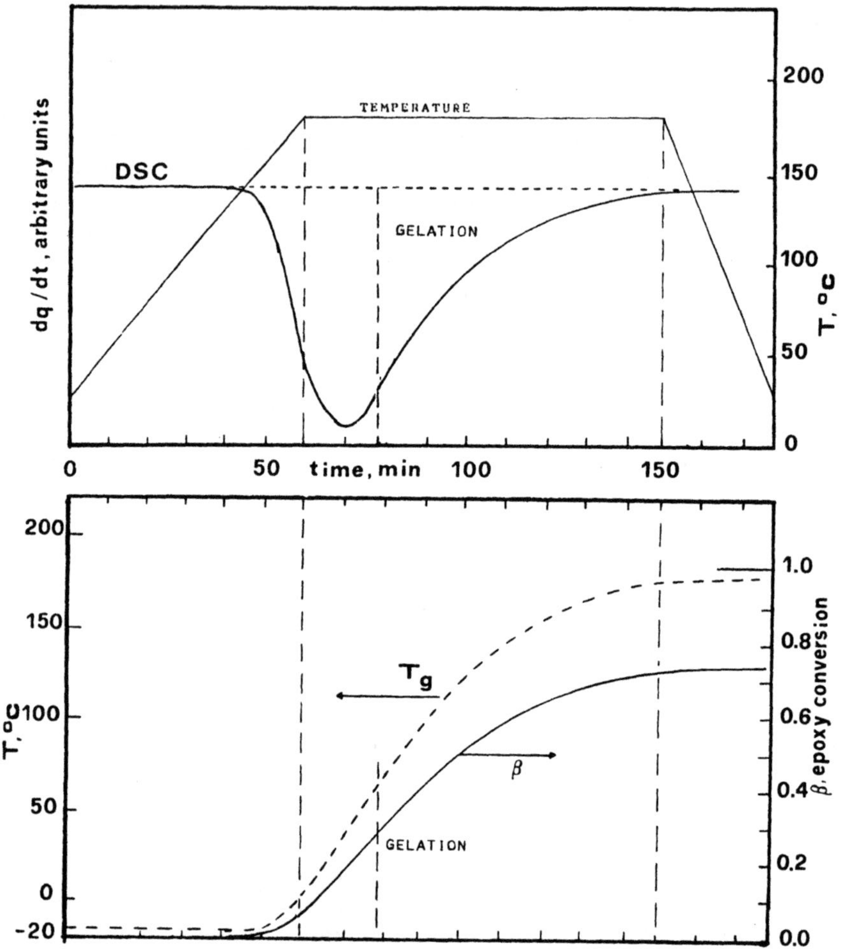

FIG. 5.2. Upper part; DSC traces during a cure cycle. Lower part; measured kinetic parameter β of epoxy conversion and changes in the system T_g.[25]

Alternatively, one of the most promising techniques for the identification of the molecular composition of epoxy systems is high performance liquid chromatography (HPLC).[22] The technique involves the use of a mobile phase pumped at high pressures, and columns packed with microparticulate stationary phase in order to separate the solute components. Sensitive detectors are used to monitor the concentration of the separated components. Separation is based upon

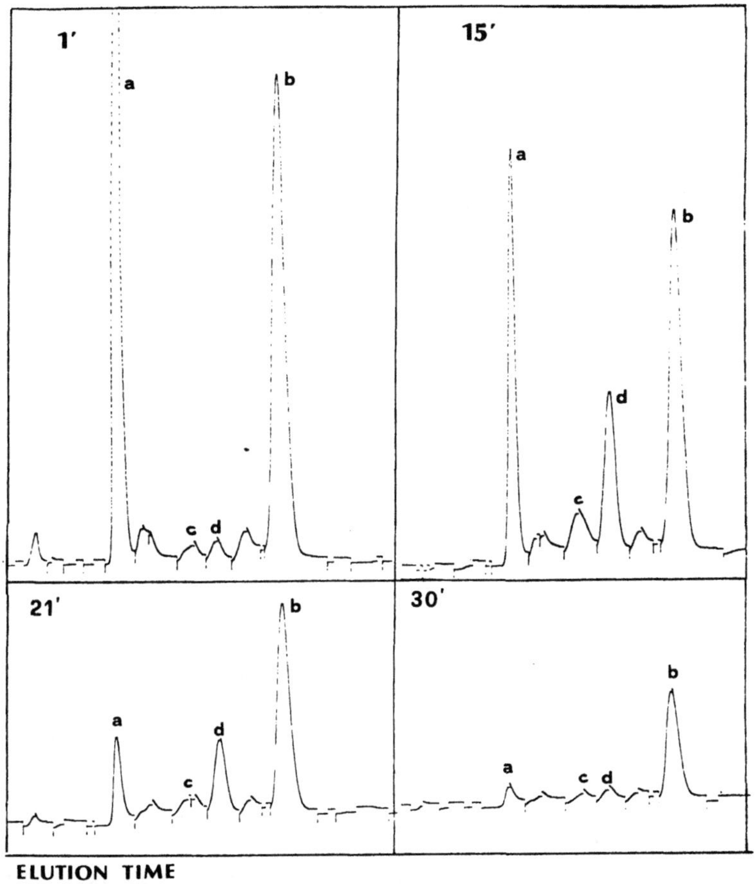

FIG. 5.3. HPLC chromatograms during isothermal cure.[23] a, DDS; b, TGDDM; c, DDS–TGDDM–DDS; d, DDS–TGDDM.

the relative solubility and distribution of the solute between the bonded and mobile phases. The intensity of the peaks in the HPLC chromatograms is proportional to the concentration of each species present in the reacting prepolymer. Figure 5.3 shows the chromatograms of a TGDDM–DDS mixture, isothermally cured at 180°C for times ranging from 1 to 35 min.[23] The concentration, expressed in weight fractions of monomers (TGDDM and DDS), has been taken as a function of time from the reduction of the intensity of the TGDDM and DDS peaks in the mixture cured for different times. The

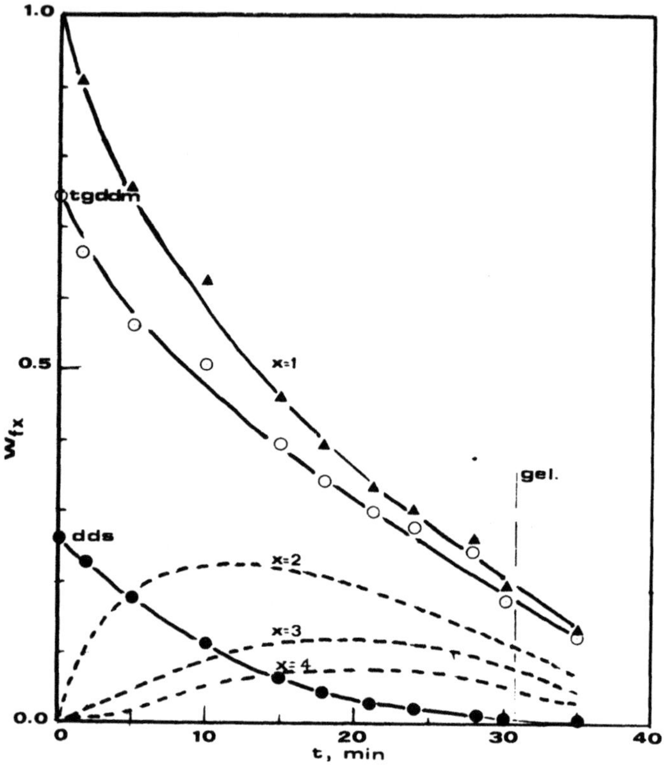

FIG. 5.4. Theoretical and experimental molecular distributions during an isothermal cure.[23] ▲, $x = 1$; ○, TGDDM; ●, DDF.

initial weight fractions of monomers were W_0(TGDDM) = 0·74 and W_0(DDS) = 0·26. The time dependency of the weight fraction of monomers and x-mers is reported in Fig. 5.4. The concentration of the species present in the system has previously been expressed in terms of the branching coefficient, α (see eqn (5)); the variation of $\alpha(t)$ has then been derived from the experimental values of the monomer concentration, and subsequently the x-mers distribution has been calculated. The actual distribution of the reaction products has been experimentally found[23] to follow the theoretical profiles of Fig. 5.4.

However, the impurities frequently present in epoxy resins, and sometimes in curing agents, may considerably affect network formation and the validity of simplified theoretical molecular modelling. Usually, the main by-products of the diglycidyl and tetraglycidyl epoxy resins are oligomers, and low amounts of reactive chlorine which, although reducing the epoxy content, increases the average functionality of the monomeric units present in the system. The variation in mean functionality strongly influences the average molecular weight, finally altering the gelation limits and the viscosity of the material during the cure. Quantitative information about the magnitude of these effects may, in principle, be employed for a correct statistical treatment.

5.3. RHEOLOGY OF THERMOSETTING REACTIONS

5.3.1. General Considerations

The processing and the final properties of thermosets depend on their composition as well as on the details of the network structure generated after the gel point. Viscosity, as indicated before, reflects the molecular size distribution and the complexity of the material and is one of the most important properties of the polymer. Accurate prediction of material properties, such as the kinetics of polymerisation and related changes in the viscosities, implies a knowledge of the basic phenomena occurring throughout the overall process.

As previously discussed, the formation of macromolecular species and subsequently of the polymer network is accompanied by a significant rise in the initial viscosity, and in the glass transition temperature of the system. Four physical states, namely liquid, ungelled glass, rubber, and gelled glass, each associated with different rheological behaviour can be observed during the polymerisation of a thermosetting resin as a function of temperature and degree of cure.

5.3.2. Constant Shear Rate and Dynamic Tests

The rheology of a liquid polymeric system during isothermal cure can adequately be described by kinetic parameters, identified in previous works[24,25] as the initial rate of increase in the viscosity, or the time needed to reach a critical molecular weight, and by the Williams–Landel–Ferry equation. The viscosity profiles for a TGDDM–DDS reactive mixture obtained[24] in steady shear during the isothermal cure at temperatures higher than the onset one (120°C) are reported in Fig. 5.5. The exposure of this mixture to high temperatures promotes the nucleophilic addition of DDS primary amines to the epoxide rings of the TGDDM. The increase in molecular weight as a consequence of

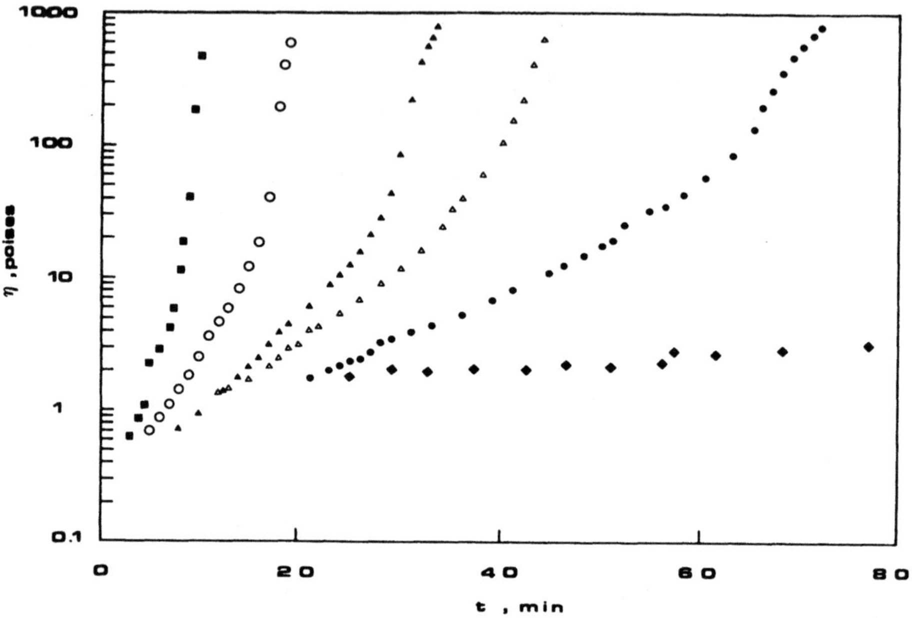

FIG. 5.5. Viscosity profiles from steady shear measurements during isothermal cure. ◆, 120°C; ●, 145°C; △, 155°C; ▲, 158°C; ○, 171°C; ■, 182°C.

the cure reaction causes an increase in the viscosity, which initially varies linearly with time. However, while the polymerisation proceeds, branching of the linear products occurs, leading to complex intermolecular interactions, such as entanglements, which further increase the viscosity of the system. An upturn is then observed before the occurrence of gelation where the viscosity measured in steady shear reaches an infinite value. The weight average degree of polymerisation

$$\bar{X}_w = (1 + \alpha)/(1 - (f - 1)\alpha) \tag{6}$$

increases rapidly as the reaction proceeds, reaching infinity at the gel point. The progressive increase in viscosity of the liquid reacting system, up to an infinite value as the gel point is approached, is explained by its dependency on weight average molecular weight.

At each temperature, the viscosity curve is characterised by a couple of critical values of time and viscosity, corresponding to the upturn critical point, as well as by the initial viscosity of the unreacted material, evaluated by extrapolating to time 'zero' the initial linear portion of the viscosity curve. While η_0 is representative of the unreacted system, the critical viscosity η_c characterises a given extent of growth of the polymer chains, and a given extent of entanglements. At any temperature, after a period of time equal to the critical time t_c, the growth of polymer chains is such that further addition of units will cause significant increases in the viscosity. As one would expect, both critical viscosity and time decrease with increasing temperature. If the values of the actual viscosity, as detected during the test, and of the actual time are normalised to the critical values, a single generalised curve, such as that reported in Fig. 5.6, is obtained. The fact that for this system, the viscosity profiles during the isothermal cure are

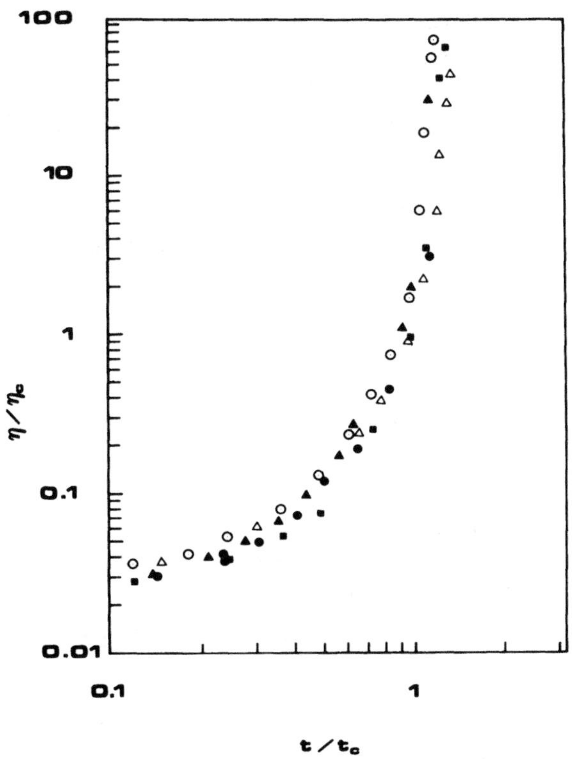

FIG. 5.6. Normalised curve of viscosity profiles measured at different temperatures.[24] ◆, 120°C; ●, 145°C; △, 155°C; ▲, 158°C; ○, 171°C; ■, 182°C.

described by a single generalised curve, suggests that at any temperature the same mechanism of reaction is active. Therefore, when the temperature is changed, although the kinetics of chain growth and their mobility are changed, the steric arrangement of the macromolecules does not vary. The cure reactions of these epoxy based resins have been extensively investigated via calorimetric and analytical methods. In particular, previous investigations have demonstrated that the principal reactions dominating the cure of the TGDDM–DDS systems are the primary amine addition before the gel point, and epoxide etherification during the crosslinking process.[13,14]

Although the liquid state can be characterised by steady-state viscosity measurements dynamic mechanical techniques are preferred for thermosets because of the extreme rheological changes occurring during the conversion of the liquid prepolymer into a solid crosslinked polymer. These changes cannot be monitored by a single mechanical test. Several dynamic mechanical techniques have been proposed[26–28] to study the reactive systems throughout the change from thermosetting liquid to solid.

Oscillatory tests are generally characterised by a cyclic deformation (usually harmonic oscillations of various frequencies) of a polymer specimen. If an homogeneous isotropic specimen is stressed according to the law

$$\tau = \tau_0 \exp(i\omega t) \tag{7}$$

where τ_0 is the intensity of the shear stress and ω the frequency, a strain is caused. The latter changes according to a periodic law of frequency ω, and generally lags in phase behind the stress by an angle δ

$$\gamma = \gamma_0 \exp(i\omega t - \delta) \tag{8}$$

Of fundamental interest for quantitative characterisation of the viscoelastic behaviour are the ratios of τ to γ, the dynamic modulus, G^*; and of τ to $\dot{\gamma}$, the dynamic viscosity η^*

$$\begin{aligned} G^* &= \tau/\gamma = \tau_0/\gamma_0 \cos\delta + i\tau_0/\gamma_0 \sin\delta \\ \eta^* &= \tau/\dot{\gamma} = \tau_0/\omega\gamma_0 \sin\delta + i\tau_0/\omega\gamma_0 \cos\delta \end{aligned} \tag{9}$$

The above complex quantities can be represented as the sums of real and imaginary components

$$\begin{aligned} G' &= (\tau_0/\gamma_0)\cos\delta & G'' &= (\tau_0/\gamma_0)\sin\delta \\ \eta' &= (\tau_0/\omega\gamma_0)\sin\delta = G''/\omega & \eta'' &= (\tau_0/\omega\gamma_0)\cos\delta = G'/\omega \end{aligned} \tag{10}$$

where G' is the storage modulus, which determines the elastic part of the stress, and G'' is the loss modulus, which determines the viscous part of the stress. The viscoelastic properties of a material can then be identified by two quantities; i.e. the ratio of the stress and strain (τ_0/γ_0) and the phase lag (δ). The particular case of $\delta = 0$, with no phase lag between stress and strain, is characteristic of an elastic medium, while the case of $\delta = \pi/2$, with the rate of strain coincident in phase with the stress, is characteristic of a perfectly viscous fluid. The loss tangent, $\tan\delta$, is usually measured in dynamic tests, and according to the previous formula, is $\tan\delta = G''/G'$.

The above quantities should not depend on the absolute values of the stress and strain in the case of linear viscoelastic behaviour, and do not apply in the case of non-linear behaviour.

Finally, when oscillatory strains are applied instead of stresses, the dynamic compliance, the reciprocal of G^*, is used in the above expressions.

First, therefore, linear viscoelastic behaviour preceding the gelation of epoxy resin reactive mixtures should be verified using dynamic tests. Sinusoidal strains and frequencies of different intensity, applied to the system considered here, confirmed the linear viscoelastic behaviour before gelation.[24] The reactive mixture, tested at temperatures where the kinetics of the cure reactions did not influence the actual molecular distribution (below 120°C), exhibits linear viscoelastic behaviour throughout the ranges of frequency and amplitude investigated.

The complex viscosity, and the ratio of the two components of the complex shear modulus measured in an isothermal cure at 180°C, are reported in Fig. 5.7. The viscous component of the complex viscosity is initially much higher than the elastic one (see vectors in Fig. 5.7) and the system behaves as a liquid. However, as its viscosity increases, molecular branching becomes significant, increasing the elastic character of the mixture. A first inflexion in the viscosity curve, which in the previous description[24] of the steady shear tests has been designated as a critical point, is then observed. The value of $\tan\delta$, initially high, suddenly decreases as the polymerisation and branching proceeds beyond the critical point, where its value is unity. A second inflexion, in the complex viscosity, occurring when $\tan\delta$ reaches a minimum, marks the formation of an infinite elastic network (gel point). The elastic component of the viscosity (η'' in Fig. 5.7) is, in fact, a maximum at this point. Other authors[29] have identified the gel point as the point where the two components of the complex shear modulus

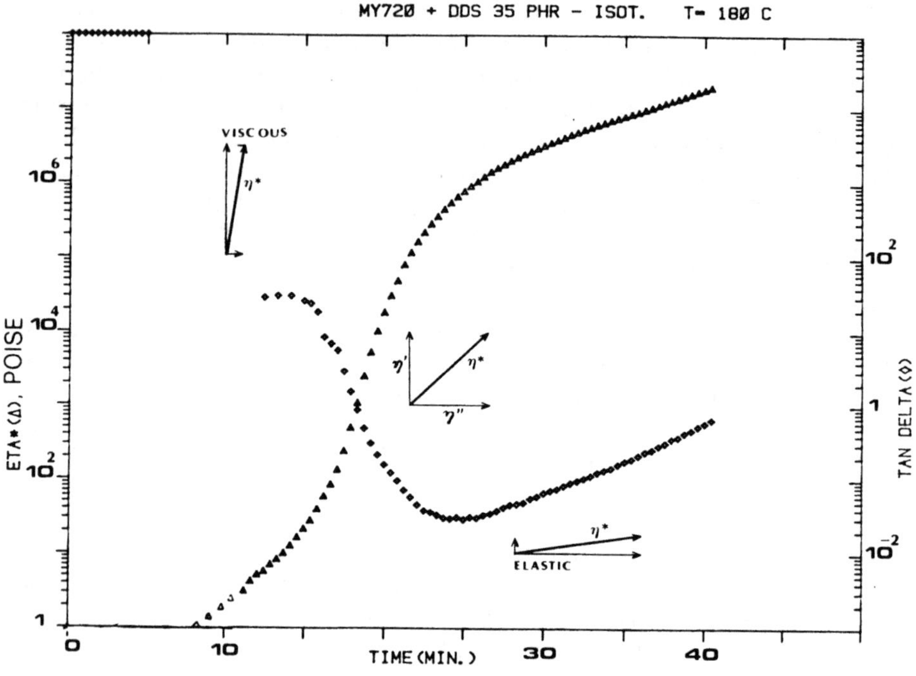

FIG. 5.7. Complex viscosity, eta*, and loss tangent, tan delta, for an isothermal cure at 180°C of a TGDDM–DDS mixture. The elastic, η'', and the viscous, η', components are qualitatively indicated at different degree of cure.

cross over ($\tan \delta = 1$). Recently, however, Han *et al.*[30,31] indicated that in an isothermal cure, the time needed to reach the maximum G' was coincident with the gelation time independently determined in steady shear flow measurements, while the system was still a viscous liquid and $\tan \delta$ was unity.

Babayesky and Gillham[32] described three different types of dynamic mechanical behaviour in the isothermal cure of epoxy–diamine systems at different temperatures. Dynamic mechanical spectra showing sample rigidity (modulus) and damping, as obtained in torsional braid analysis (TBA), and reported in Fig. 5.8, were used. They described a first kind of behaviour at low temperatures, where the sigmoidal increase in the rigidity (proportional to the storage modulus) was accompanied by a single damping peak or, at higher temperatures, also by a damping shoulder. In the second type of behaviour this shoulder resolves into a second damping peak, while the relative rigidity shows two distinct increases corresponding with the two damping peaks. Finally, the third kind of behaviour was characterised by the disappearance of the first damping peak. These authors recognised that the first damping peak was associated with the gel point, while the second was associated with the transition to the glassy

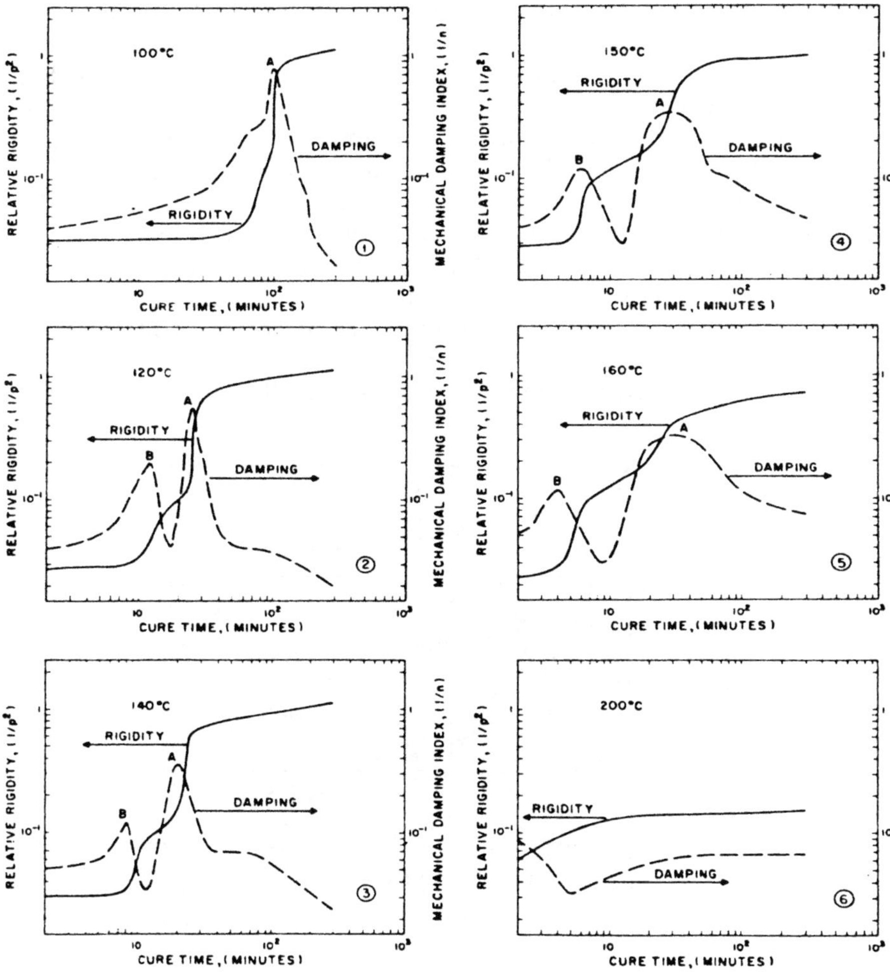

FIG. 5.8. Rigidity and mechanical damping vs time during isothermal cure of an epoxy resin.[32]

state. The increase in the molecular weight causes the vitrification to occur before gelation in the first low temperature behaviour. In the second type of dynamic mechanical response, the reactive system first passes through gelation, giving the second damping peak (B in Fig. 5.8), and then vitrifies giving rise to the first peak (A in Fig. 5.8). If the cure temperature is sufficiently high, exceeding the maximum glass transition temperature of the completely crosslinked system, only the peak associated with the gelation is observed (third type of behaviour).

This behaviour is not unique to epoxy–aromatic amine systems, but is typical of most thermosets. A general time temperature transformation diagram (TTT), proposed by Gillham *et al.* (see, for instance, Ref. 33), can be obtained from the dynamic mechanical characterisation previously described. These useful plots, Fig. 5.9, describe the state of a curing thermoset as a function of temperature and time. In these TTT diagrams several different zones and characteristic temperatures and times, associated with different chemo-rheological behaviour, can be observed. Cure reactions do not occur at temperatures lower than the glass transition temperature (T_g) of the initial liquid resin (this range of temperatures is usually used to store the

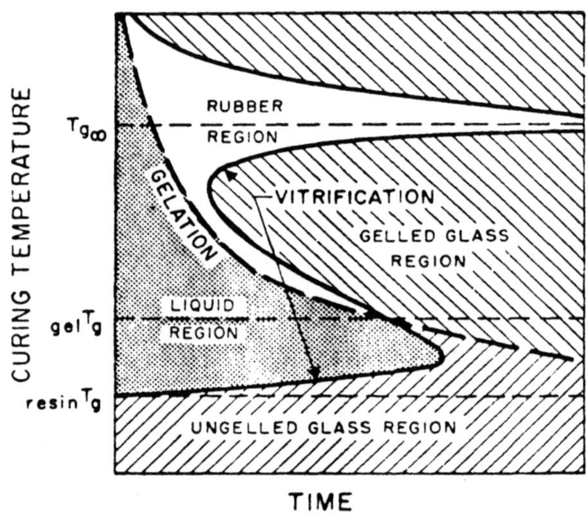

FIG. 5.9. Time–temperature transformation diagram for thermosetting resins.[33]

reactive prepolymer). At temperatures above T_{gg} (the temperature at which the times to gelation and vitrification are the same) gelation precedes vitrification. Between T_g and T_{gg}, the system will react until its glass transition temperature exceeds the cure temperature, quenching the reactions. Above T_g the system gels but does not vitrify. While the time to gelation is expected to decrease with increasing temperature, the time to vitrification is observed to pass through a minimum, occurring at intermediate cure temperatures. This is because of competition between the increased reaction rate constants and the increased degree of cure needed to overcome the thermal motions for vitrification.

5.3.3. Viscosity Modelling

Several authors have considered theoretically the rheological behaviour of a curing thermoset through proposed models[25,34–36] of different levels of complexity. In particular, May *et al.*[35] illustrated the feasibility of a model containing four parameters, subsequently expanded to six, which was able to describe the rheology of particular epoxy–amine systems. The 'four parameter' model relates the viscosity of a particular resin system to four material parameters, and to time and temperature. Starting from the consideration of Roller[37] that the viscosity could be expressed as a function of the melting behaviour ($\eta(T)$) and of the kinetics of the reaction (φkt)

$$\eta(T, t) = \eta(T) \exp(\varphi kt) \tag{11}$$

where k is an apparent kinetic factor, and φ a function relating the extent of reaction and viscosity, May *et al.*[35] derived the expression

$$\ln \eta(T, t) = \ln \eta_x + (E_\eta/RT) + \varphi_{kx} \int \exp(E_x/RT)\, dt \tag{12}$$

This model contains four parameters characteristic of each reacting system, namely: E_η, the flow activation energy; η_x, the pre-exponential viscosity; φ_{kx}, the pre-exponential kinetic constant; and E_x, the reaction activation energy. The isothermal viscosity profiles, such as those presented in Fig. 5.5, can be used to determine the initial viscosity, $\eta_0(T)$ and the initial slopes, $k(T)$ (proportional to the actual kinetic factor) at different temperatures. Arrhenius plots of these quantities lead to the identification of the characteristic flow and reaction activation energies, E_η and E_x, and the corresponding pre-exponential factors, η_x and k_x. The principal assumptions inherent in this model, which can seriously limit the validity of its predictions, concern the values of η_x, which is considered independent of the molecular weight, and the flow activation energy, E_η, which, although considered constant here, has been reported by Kamal *et al.*[38] to vary significantly during the pre-gelation polymerisation. A further over-simplification concerns the overall reaction order, assumed to be unity. A second six-parameter model was derived, treating the overall reaction order and chain entanglements as separate, independent parameters.[39]

Alternative theoretical approaches, which explicitly consider the molecular and temperature dependence of the viscosity have been recently presented.[23,34] The rheological behaviour of a reacting system is, in fact, principally governed by two effects, one related to the structural changes caused by the cure reactions (i.e. the chemical aspect), and the other to the variation in molecular mobility induced by temperature variations (i.e. the physical aspect). The dependence of viscosity on temperature may be described by using the William, Landel and Ferry (WLF) equation.[40]

$$\ln(\eta/\eta_r) = -C_1(T - T_g)/((C_2 + T - T_g) + C_1(T_r - T_g)/(C_2 + T_r - T_g) \tag{13}$$

where η_r is the viscosity at the reference temperature T_r, C_1 and C_2 are constants, and T_g is the glass transition temperature of the unreacted system, or of any intermediate state the system passes through to reach the critical stage and gelation.

Figure 5.10 reports the prediction capability of the WLF equation for a TGDDM–DDS system in the unreacted state. A range of temperatures where the reactions do not occur was used. At high temperatures, the extrapolated values of the initial viscosity were employed.[24]

On the other hand, any variation in viscosity is determined, at a fixed temperature, by the changes in molecular characteristics and may be derived using the following assumptions. For a linear polymer, the viscosity is related to the molecular weight by the equation[41]

$$\eta = KM_w^a \tag{14}$$

where $1 < a < 2{\cdot}5$ for M_w lower than a critical value, and $a = 3{\cdot}4$ for M_w higher than the critical value. A similar relationship has been proposed by Valles and Macosko[42] for a non-linear copolymerisation

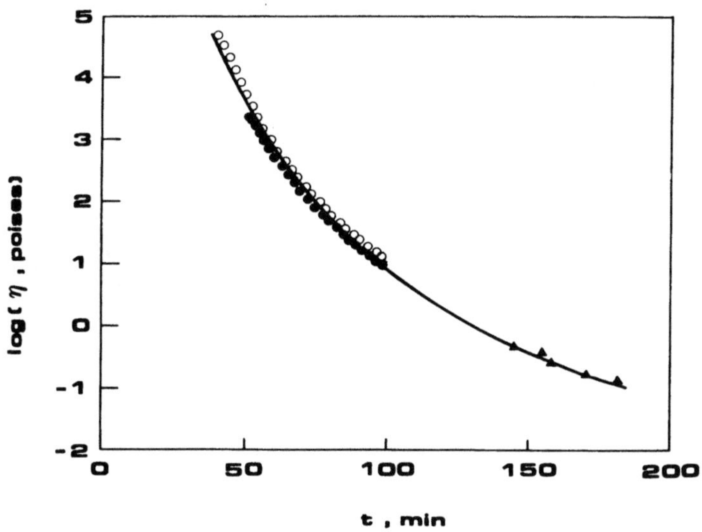

FIG. 5.10. Experimental and theoretical (WLF, eqn (13)) profiles of the viscosity for an unreacting epoxy system. Dynamic mode at 1 (○) and 10 (●) Hz, and extrapolated values in isothermal cure of Fig. 5.5.[24]

such as that considered here

$$\eta = k(gM_w)^a \tag{15}$$

with $a = 3{\cdot}4$ and g, the ratio of the square of the radii of gyration of a branched to a linear chain of the same molecular weight, defined as[43]

$$g = ((1 - \alpha(f - 1))/2\alpha) \ln ((1 - \alpha)/(1 - \alpha(f - 1)) \tag{16}$$

If we normalise the viscosity to its initial value at $\alpha = 0$, then

$$\eta(\alpha)/\eta_0 = (gM_w(\alpha)/M_{w_0})^{3{\cdot}4} = (g\bar{X}_w)^{3{\cdot}4} \tag{17}$$

where $\bar{X}_w$ is the number average molecular weight previously defined (eqn (6)).

A combination of the previous expressions lead to the relationship

$$\eta(\alpha) = \eta_0 \left[\frac{1-\alpha}{2\alpha} \ln \frac{1-\alpha}{1-\alpha(f-1)}\right]^{3{\cdot}4} \tag{18}$$

The experimental viscosity profile in an isothermal test carried out at 180°C is adequately described by eqn (18), as illustrated in Fig. 5.11.

The viscosity changes in a reactive system continuously heated up to

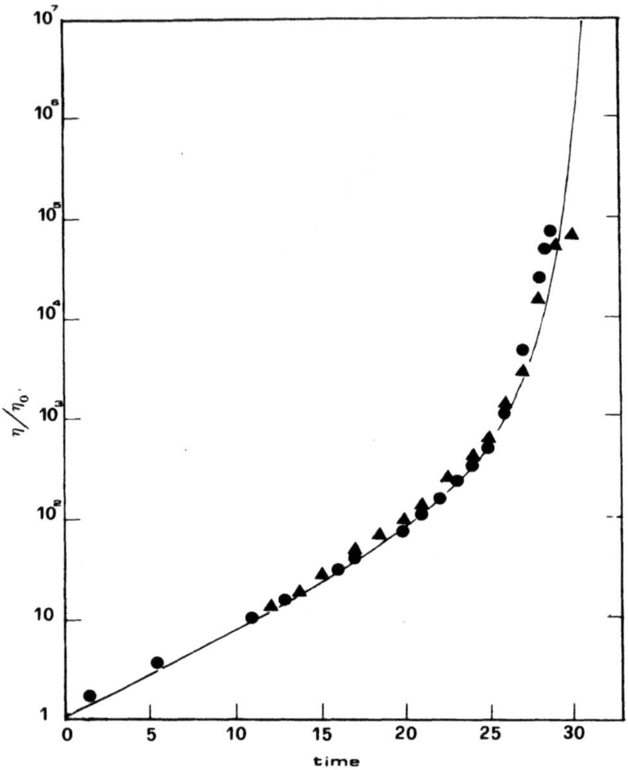

FIG. 5.11. Experimental (▲, dynamic shear; ●, constant shear) and theoretical (eqn (18)) viscosity profiles in an isothermal cure.[23] $\eta/\eta_0 = [(1+\alpha_t)/(1-\alpha_t(f-1))]^{3\cdot4}$.

the final cure temperature should be predicted by combining the expressions containing the temperature (eqn (13)) and the molecular dependency (eqn (18)),

$$\eta(T, \alpha) = \eta_{r0}\left[\frac{1-\alpha}{2\alpha}\ln\frac{1-\alpha}{1-\alpha(f-1)}\right]^{3\cdot4} \times \frac{\exp C_1(T_r - T_{g0})/(C_2 + T_r - T_{g0})}{\exp C_1(T - T_g(\alpha))/(C_2 + T - T_g(\alpha))} \quad (19)$$

where η_{r0} is the viscosity of the unreacted system at a reference temperature and T_{g0} is its glass transition temperature.

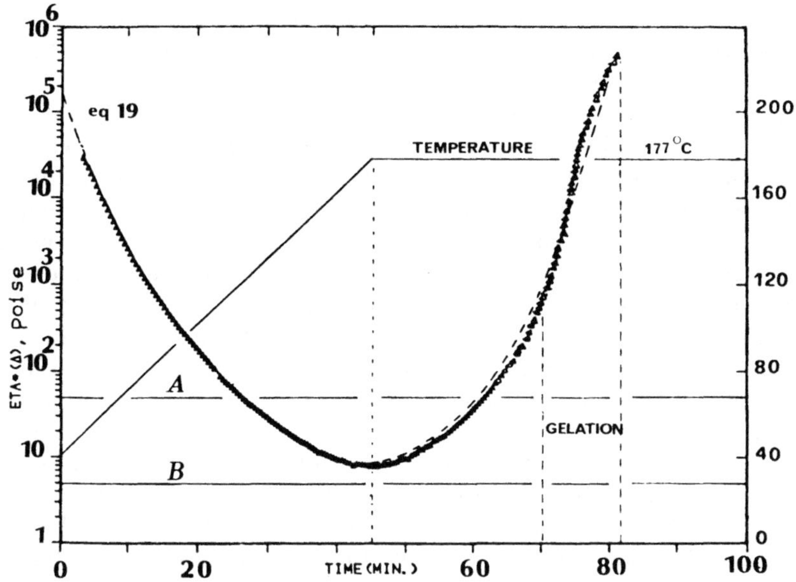

FIG. 5.12. Experimental (△) and theoretical (eqn (19)) viscosity profiles during a cure cycle. Gelation (– – –) and viscosity (——) limits (see text).[25]

The rate of variation of the glass transition temperature with the degree of cure has been determined by differential scanning calorimetry, and is illustrated in Fig. 5.2 (lower part). The data are used in eqn (19). Theoretical studies, however, are applied to relate T_g to molecular parameters.[44] The predictability of eqn (19) is successfully tested in Fig. 5.12.

5.4. CURE CYCLE AND MATERIAL CONTROL

Viscosity control during polymer processing is particularly critical for thermosetting systems, since viscosity may vary with temperature, flow conditions and time, due to the chemical reactions occurring in the liquid state. Chemoviscosity is a term which indicates the variations in the viscosity induced by the chemical reactions, and chemorheology is the science of the viscoelastic behaviour of the reacting system. This knowledge is very important in the fabrication of polymer composite parts, especially for aerospace applications, where precise resin cure control is required. The viscosity–time profiles must allow for simultaneous bubbling-off of trapped gas and sorbed moisture, for fibre compaction, and for resin flow before the resin has gelled and ceases to flow. The polymerisation kinetics and chemoviscosity should therefore, be controlled by appropriate temperature programming.

The principal physical events occurring during a cure cycle were briefly discussed in the previous sections and are illustrated in Fig. 5.12. Thus, the viscosity initially decreases with time, as the temperature is increased, but the reactions are not yet activated. During this first stage, sorbed volatiles should start to diffuse out of the resin. When the viscosity reaches a minimum and then begins to increase, as a result of the thermosetting reactions, trapped bubbles must be allowed out of the composite. Bubbles, in fact, become mobilised, when the viscous flow forces are sufficiently large to overcome the surface tension forces.[45] Pressure, which induces resin flow, should be correctly applied in a limited viscosity range in order to remove the excess resin and trapped bubbles properly and to consolidate the plies. Figure 5.12 shows qualitatively the viscosity limits (A and B) that must be matched by the resin. The lower limit (B) is imposed by the flow characteristics of the system, to avoid an excessive loss of resin from the composite, and to ensure the flow forces necessary to mobilise the bubbles. Too low a viscosity will produce a composite with insufficient resin, and with trapped bubbles. Too high a viscosity, on the other hand, does not allow a sufficient flow of resin and compaction of the system. Bubbles, again, are not removed from the composite. Other important physical events, such as void nucleation and volatile diffusion and desorption, must be considered, and have been analysed by other authors.[45]

The analysis and management of curing is, hence, one of the major areas of concern for the reliable development of composite materials. Generally, studies such as the TTT diagrams may be useful in the early investigations and interpretations of the complex interactions of the changing chemical structure and the rheological state during cure, but new experimental techniques are needed to control the process continuously. Viscosity, which is one of the most important process variables, should be monitored and controlled to ensure the consistent and reliable production of fibre reinforced composites by autoclave processes. No techniques are currently available for the direct measurement of the viscosity during the cure. Physical restrictions, in fact, prevent the use of the traditional tools of investigation such as calorimetry, chromatography or infrared spectroscopy. Recently, however, several methods based on the measurement of dielectric properties have been developed[46–48] and proposed for the indirect measurement of viscosity changes.[49] May *et al.*[49] have shown that the dielectric properties of a curing resin correlate with the chemical and rheological changes. The use of dielectric signals permits the development of process control from a closed-loop automated cycle.

5.5. CONCLUDING REMARKS

The cure of epoxy resins is an example of thermosetting reactions in which the glass transition of the reacting material often becomes higher than the reaction temperature, owing to diffusion control before all functional groups have reacted. The mobility, and thus the apparent reactivity, of the functional groups may not decrease to the same extent, since they depend on the size and steric arrangement of the molecule to which the group is attached. If the reactions of less hindered groups are preferred, the degree of polymerisation distribution would become narrower, and the critical conversion at the gel point would be shifted to different values, depending on the glass transition temperature reached by the material. As a direct consequence of diffusion constraints, the network formed could have a different structure, even if the same conversion is reached.

The considerably lower tensile strengths of thermosets than those predicted[50] by primary or van der Waals bond considerations have been attributed to the presence of morphological inhomogeneities,[51,52] theoretically related to differences in crosslinking densities.[53,54] However, other authors[55,56] have found no evidence of inhomogeneous crosslinking, at least for simple epoxy systems composed of compatible units, where alternating stepwise polyaddition controls the network formation. Dusek *et al.* in a series of papers[56–58] identified monomer functionality, the relative reactivities of the functional groups participating in the cure, initial composition, and the reaction mechanisms as the factors affecting the network formation and structure of epoxy resins. Such conclusions, however, do not generally apply to all epoxy systems. The occurrence of parallel reactions, as in the case of the epoxy–amine addition and subsequent polyetherification, and in systems formed by less compatible components (e.g. in presence of impurities which also lower the reactivity of the resin), may induce pronounced fluctuations in crosslinking density. In the addition with polyamines, the epoxy group has been considered monofunctional, but becomes bifunctional if the hydroxyl formed reacts with another epoxy group (polyetherification). Although in ordinary temperature conditions the polyetherification reaction rate has been reported to be an order of magnitude smaller than for amine epoxide addition,[14] polyetherification can be more favoured in presence of a basic tertiary nitrogen atom.[14,59] Modulus asymmetry and preferential swelling may be induced by the occurrence of polyetherification.[60,61] A nonhomogeneous structure has been hypothesised to explain anomalies in the sorption behaviour and ageing characteristics of TGDDM–DDS cured resins under thermal cycling conditions.[61]

The topography of crosslinked networks generated in a thermosetting epoxy resin by the cure reaction strongly depend on the cure thermal history and on the system's initial composition. An approach to structure–property relationships needs an investigation of the system's chemorheology and thermokinetics, and correlation of the properties of the cured material with variations in composition and with the initial state of the prepolymer.

REFERENCES

1. FLORY, P. J., *Principles of Polymer Chemistry,* Cornell University, Ithaca, New York, 1953.
2. APICELLA, A. and NICOLAIS, L., *ACS, Ind. Eng. Chem. Prod. Res. Dev.,* **23** (2), 288 (1984).
3. MORGAN, R. J., O'NEAL, J. and MILLER, D. B., *J. Mater. Sci.,* **14,** 109 (1979).
4. SCHNEIDER, N. S., SPROUSE, J. F., HAGNOUER, G. L. and GILLHAM, J. K., *Polymer Eng. and Sci.,* **19,** 304 (1979).
5. BELL, J. P., *J. Polym. Sci.,* **A2,** 417 (1970).
6. DUSEK, K., *Macromol. Chem. Suppl.,* **2,** 35 (1979).
7. GORDON, M., *Proc. Roy. Soc.,* **A268,** 240 (1962).
8. BUCHARD, W., *Adv. Polym. Sci.,* **48,** 1 (1982).
9. MACOSKO, C. W. and MILLER, D. R., *Macromolecules,* **9,** 199 (1976).
10. CAROTHERS, W. H., *Trans. Farad. Soc.,* **32,** 39 (1936).
11. CAROTHERS, W. H., *J. Am. Chem. Soc.,* **51,** 2548 (1929).
12. DUSEK, K., SCHECHTER, L., WYNSTRA, J. and KUNKJY, R. P., *Ind. Eng. Chem.,* **48,** 94 (1956).
13. MOACANIN, J., CIZMECIOGLU, M., TSAI, F. and GUPTA, A., *Am. Chem. Soc., Org. Coat., Appl. Polym. Sci. Proc.,* **47,** 587 (1982).
14. GUPTA, A., CIZMECIOGLU, M., COULTER, D. and LIANG, R. H., *J. Appl. Polym. Sci.,* **28,** 1011 (1982).
15. STARK, E. B., SEFERIS, J. C., APICELLA, A. and NICOLAIS, L., *Therm. Acta,* **77,** 19 (1984).

16. Apicella, A., Nicolais, L., Iannone, M. and Passerini, P., *J. Appl. Polym. Sci.*, **29** (6), 2083 (1984).
17. Flory, P. J., *J. Am. Chem. Soc.*, **63**, 3091 (1941).
18. Stockmayer, W. A., *J. Chem. Phys.*, **11**, 45 (1954).
19. Prime, R. B., *Polym. Eng. Sci.*, **13** (5), 365 (1973).
20. Roger, R. N. and Smith, L. C., *Therm. Acta*, **1**, 1 (1970).
21. Dusek, K., Plestie, J., Lendnicky, F. and Lunak, J., *Polymer*, **19**, 393 (1978).
22. Hagnauer, G. L. and Setton, I., *J. Liquid Chromatogr.*, **1**, 55 (1978).
23. Halpin, J. C., Apicella, A. and Nicolais, L., Processing of thermosets, in *Polymer Processing and Properties*, Eds, G. Astarita and L. Nicolais, Plenum Press, New York, 1984.
24. Apicella, A., Masi, P. and Nicolais, L., *Rheol. Acta*, **23**, 291 (1984).
25. Apicella, A., Nicolais, L. and Halpin, J. C., The role of the processing chemo-rheology on the ageing behaviour of high performance epoxy composites, in *Proceedings of the 28th National SAMPE, Materials and Process Continuing Innovations*, Anaheim (USA), 1983.
26. Lewis, A. F. and Gillham, J. K., *J. Appl. Polym. Sci.*, **6**, 422 (1962).
27. Gillham, J. K., *CRC Critical Reviews in Macromolecular Science*, **11** (4), 295 (1972).
28. Gordon, M. and Grieveson, B. M., *J. Polym. Sci.*, **29**, 9 (1958).
29. Tung, C. M. and Williams, H. L., *Polym. Eng. Sci.*, **22**, 254 (1982).
30. Han, C. D. and Lem, K. W., *ACS Org. Coat. Appl. Polym. Sci. Proc.*, **47**, 581 (1982).
31. Yap, C. Y. and Williams, H. L., *Polym. Eng. Sci.*, **22**, 254 (1982).
32. Babayesky, P. G. and Gillham, J. K., *J. Appl. Polym. Sci.*, **17**, 2067 (1973).
33. Gillham, J. K., Torsional braid analysis of polymers, in *Developments in Polymer Characterisation—3*, Ed., J. V. Dawkins, Applied Science Publishers, London, 1982, Ch. 5.
34. Tajima, Y. A. and Croizer, D., *Polym. Eng. Sci.*, **23** (4), 186 (1983).
35. Wereta, A., Dusi, M. R. and Hadad, D. K., *ACS Chem. & Ind. Eng. Symp.*, New York, August 1981.
36. White, R. P., *Polym. Eng. Sci.*, **14** (1), 50 (1974).
37. Roller, M. B., *Polym. Eng. Sci.*, **15** (6), 406 (1975).
38. Kamal, M. R., Sourour, S. and Rayan, M., *SPE Tech. Conf.*, Montreal, 1973.
39. Dusi, M. R., May, C. A. and Seferis, J. C., *ACS Org. Coat. Appl. Polym. Sci. Proc.*, **47**, 635 (1982).
40. Ferry, J. D., *Viscoelastic Properties of Polymers*, Wiley Interscience, New York, 1971, Ch. 20.
41. Berry, G. C. and Fox, T. G., *Adv. Polym. Sci.*, **5**, 261 (1968).
42. Valles, E. M. and Macosko, C. W., *Macromolecules*, **22**, 521 (1979).
43. Zimm, B. H. and Stockmayer, W. H., *J. Chem. Phys.*, **17**, 1301 (1949).
44. Nielsen, L. E., *J. Macromol. Sci. Rev., Macromol. Chem.*, **C3**, 69 (1963).
45. Sheidegger, A. E., *The Physics of Flow Through Porous Media*, 3rd Edition, University of Toronto Press, 1979.

46. Crabtree, D. J., *Am. Chem. Soc., Div. Polym. Chem.*, **22** (2), 233 (1981).
47. Wereta, A. and May, C. A., *J. Adhesion*, **12,** 317 (1981).
48. Valoff, S. A. and Wrasidlo, W. J., *J. Appl. Polym. Sci.*, **16,** 2159 (1972).
49. May, C. A., Dusi, M. R., Fritzen, J. S., Hadad, D. K., Maximovich, M. G. and Thrasher, K. S., *Appl. Polym. Sci. Proc.*, **47,** 419 (1982).
50. deBoer, J. H., *Trans. Faraday Soc.*, **36,** 10 (1936).
51. Blokland, R. and Prins, W., *J. Polym. Sci.*, **7** (A2), 1595 (1962).
52. Cuthrell, R. E., *J. Appl. Polym. Sci.*, **11,** 1495 (1967).
53. Solomon, D. H., Loft, B. C. and Swift, D. J., *J. Appl. Polym. Sci.*, **11,** 1593 (1967).
54. Cuthrell, R. E., *J. Appl. Polym. Sci.*, **12,** 1263 (1968).
55. Lunak, S., Vladyka, J. and Dusek, K., *Polymer*, **19,** 931 (1978).
56. Dusek, K., Ilavsky, M. and Lunak, S., *J. Polym. Sci., Polym. Symp.*, **No. 53,** 29 (1975).
57. Dusek, K., Bleha, M. and Lunak, S., *J. Polym. Sci., Polym. Chem. Ed.*, **15,** 2393 (1977).
58. Dusek, K. and Ivlasky, M., *Colloid Polym. Sci.*, **28,** 605 (1980).
59. Mones, E. T. and Morgan, R. J., *Polymer Preprints*, **22** (2), 249 (1971).
60. Chang, T. D., Carr, S. H. and Brittain, J. O., *Polym. Eng. Sci.*, **22,** 1213 (1982).
61. Apicella, A. and Nicolais, L., *Ind. Eng. Chem. Prod. Res. Dev.*, **23,** 288 (1984).

Chapter 6

ELECTRICAL METHODS OF CHARACTERISING CURE PROCESSES IN RESINS

DAVID E. KRANBUEHL

Department of Chemistry, College of William and Mary, Williamsburg, Virginia, USA

SUMMARY

The major objective of research on electrical methods of characterising cure processes in resins is to develop on-line electrical instrumentation for non-destructive material evaluation, for determination of fabrication cycle window boundaries of thermosets and thermoplastics, and for on-line closed loop cure cycle control. The key to achieving this goal is to relate the chemistry of the cure cycle process to the dielectric properties of the polymer system by correlating the time, temperature and frequency dependence of electrical measurements with chemical characterisation measurements, i.e. viscosity, ultrasonics and differential scanning calorimetry. Measurements of geometry independent variables such as the complex permittivity (ε^*) *need to be made rapidly over a wide frequency range. In turn, the time–temperature and frequency dependence of the electrical measurements provide the basis for interpreting and understanding the signal in terms of the chemistry and physics of the resin system. Dynamic electrical measurements, together with chemical characterisation work can then be used to determine: resin quality, composition and age; cure cycle window boundaries; onset of flow and point of maximum flow; extent of reaction and completion of reaction; evolution of volatiles;* T_g*; crosslinking and molecular weight buildup.*

In this chapter, as examples of the use of dynamic electrical measurements, the cure processes of the polyimides, LARC-160 and

PMR-15, and the DGEBA type epoxy Epon 828 cured with agent U have been reviewed. The thermoplastic resins UDEL-P1700 (a polysulphone) and PPQ (a polyphenylquinoxaline) have also been examined over a broad temperature range. The use of the magnitude of the complex permittivity (ε^*), *the specific conductivity* $\sigma(\text{ohm}^{-1}\,\text{cm}^{-1})$ *and a characteristic relaxation time* τ *to monitor the state of the material is discussed.*

6.1. INTRODUCTION

Dynamic electrical measurements made over a wide range of frequency provide a sensitive and convenient means for monitoring the cure process in thermosets and thermoplastics. Electrical measurements are of particular importance for quality control monitoring of the cure process in complex resin systems, because they represent one of only a few instrumental techniques available for studying molecular properties in both the liquid and solid states. Furthermore, it is probably the only convenient experimental technique for studying the cure process continuously, that is, to examine the cure chemistry throughout the process of going from a monomeric liquid of varying viscosity to a crosslinked, insoluble, high temperature solid.

The major long-range objective of research on electrical methods is to develop on-line electrical measurement techniques for quantitative nondestructive material evaluation and closed loop 'smart' cure cycle control. The key to achieving this goal is to relate the chemistry of the cure cycle process to the electrical properties of the polymer system by correlating the time, the temperature, and frequency dependent electrical measurements with chemical characterisation measurements. Recent dynamic electrical measurements, together with the chemical characterisation work, have been shown in our laboratory to have the potential to determine: resin quality, composition and age; cure cycle window boundaries; onset of flow and maximum point of flow; extent of and completion of reaction; evolution of volatiles; T_g; crosslinking and molecular weight buildup.[1–4] Similarly encouraging results have been observed in recent reports by Senturia,[5–6] Sanjana,[7] Baumgartner[8] and Mitchell[9] and their co-workers.

6.2. SHORTCOMINGS OF EARLIER WORK

Interest in the use of electrical measurements to characterise materials has existed for over fifty years. Many excellent books reviewing this literature have been written.[10–17] The extensive use of electrical measurements, primarily utilising dielectric techniques, to characterise materials has led to the current interest in using electrical measurements to monitor cure processes in resins. The reason for this interest is that electrical measurements are one of only a few techniques for convenient and continuous monitoring of the cure process as the resin's viscosity varies from less than 10^2 to more than 10^9 poise. Furthermore, from the frequency dependence of the electrical parameters it is possible to obtain molecular information on the chemistry and physics of the polymerisation reaction.

Until the last several years, research on the use of electrical measurements has often fallen short of expectations. While reasons for this are in most cases complex, usually they were due in part to the then available instrumentation, failure to make measurements over a wide range of frequencies, not understanding the molecular basis for the signal in terms of dipolar and ionic phenomena, and lack of an integrated approach which interprets the electrical measurements along with other chemical characterisation measurements.

The effective and successful use of dynamic electrical measurements for cure cycle monitoring requires an extensive effort, involving basic research on the chemistry of each resin, using a variety of characterisation techniques, which are then correlated with dynamic electrical measurements made over a wide range of frequencies. Only with this background information is it possible for electrical measurements to be most effectively used to monitor the cure process and to provide the input for a complete mathematical processing model for quality assurance cure cycle monitoring as well as closed loop 'smart' cure cycle control.

Overall, there are three major areas which need to be addressed to effectively use electrical measurements to monitor the resin's cure process. The first is instrumentation. Major advances have been made in this area. A number of rapid, low noise automatic bridges which span six decades or more in frequency have been introduced commercially during the past few years. The electrical properties of individual resins vary greatly, and may change by factors of 10^6 or more during the cure process. Thus, the selection of instrumentation and its components will eventually need to be designed and matched to a particular resin's properties, the use objectives, the operating conditions, plant processing constraints, software, etc. Selection of equipment, modification and design, are often less than a quarter of the research problem.

A second major area, often representing over half of the problem, is interpreting and understanding the molecular basis of the electrical signal. Critical aspects of this problem include making rapid measurements over a wide range of frequencies and then understanding and interpreting the molecular basis of the frequency dependence of the signal in terms of the chemistry and physics of the resin system. Essential to this step is extensive chemical characterisation work. In our laboratory we are relating frequency dependent electrical measurements to carbon-13 nuclear magnetic resonance, gel permeation chromatography, infra-red spectroscopy, viscosity, ultrasonics and differential scanning calorimetry data.

The third aspect of the problem is integrating the chemical and flow information embedded in the complex electrical signal's magnitude and frequency dependence into the processing model used for quality assurance and/or closed loop 'smart' cure cycle control. The development of processing models is an extensive subject, which is also currently under development.[17] It is dealt with throughout the volumes of this series.

6.3. MEASUREMENT

Measurements of the electrical properties of a material are usually made with an impedance bridge. An impedance bridge measures the opposition that the material presents to an alternating current in terms of the complex ratio of the voltage to the current.

$$Z^* = \frac{V(\omega)}{I(\omega)} \tag{1}$$

where Z^* is the complex impedance. It measures the magnitude and time shift of the voltage relative to the current. Often the properties of the material and its impedance are conveniently represented by an equivalent circuit of a capacitor (C_s) and resistor (R_s) in series. In this case

$$Z^* = R_s + \frac{1}{i\omega C_s} \tag{2}$$

Similarly, the properties of the material can be represented by an equivalent circuit of a resistor (R_p) and capacitor (C_p) in parallel, in which case

$$\frac{1}{Z^*} = \frac{I(\omega)}{V(\omega)} = Y^*$$
$$Y^* = \frac{1}{R_p} + i\omega C_p = G_p + i\omega C_p \tag{3}$$

where Y^* is the admittance and the reciprocal of the parallel resistance is the conductance G_p.

To describe the properties of a capacitor filled with a poorly conducting material a third notation is most often used. In this notation the equivalent circuit is viewed as simply a capacitor with capacitance C_0 without the material, and capacitance $C = \varepsilon C_0$ with the material. The magnitude and time shift of the current with respect to the voltage is represented by the complex value of the material's permittivity ε^* where $\varepsilon^* = \varepsilon' - i\varepsilon''$. Again, from elementary circuit theory

$$\frac{1}{Z^*} = \frac{I(\omega)}{V(\omega)} = i\omega\varepsilon^* C_0 = \varepsilon''\omega C_0 + i\omega\varepsilon' C_0 \tag{4}$$

All of these representations are simply two-parameter equivalent circuit representations, describing the propagation of an electromagnetic field through a material. Given one pair of parameters, it is possible to re-express them in terms of another equivalent circuit pair.

The electrical conductance of resins is usually low, and their capacitive properties are moderate to high. Therefore it is customary to treat the resin's electrical properties in terms of the third notation, the complex permittivity ε^*. Furthermore, since most polymeric materials are poor conductors, the term dielectric properties is often used in place of electrical properties.

6.4. EQUIPMENT

Most measurements of the electrical properties of resins are made using an impedance bridge in the frequency range of 1 Hz to 10 MHz. Microwave and optical measurement techniques are used at higher frequencies. For lower frequencies measurements are made using time-domain Fourier-transform techniques. Details of impedance analyser and time-domain techniques are given elsewhere.[1–12]

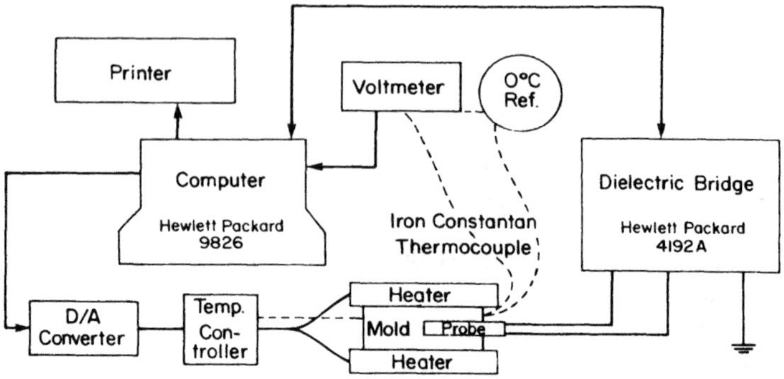

Fig. 6.1. Block diagram of apparatus used to measure the electrical properties of curing polymer systems.

An example of an automated impedance analyser apparatus is shown in Fig. 6.1. It consists of a Hewlett–Packard 4192A LF Impedance Analyser, controlled by a 9836 Hewlett–Packard computer. The resin is cured in a 3 in. (75 mm) mould placed between thermostated heating plates. The time–temperature profile of the mould is controlled by a program in the computer. The lay-up consists of a number of layers of Kapton® to insulate the probe from the metal mould, a probe which can be inserted directly between layers of polymer resin and further layers of Kapton between this and the mould top (Fig. 6.2). An iron–Constantan thermocouple is attached directly to the mould and the temperature is measured by a Keithley 179 TRMS

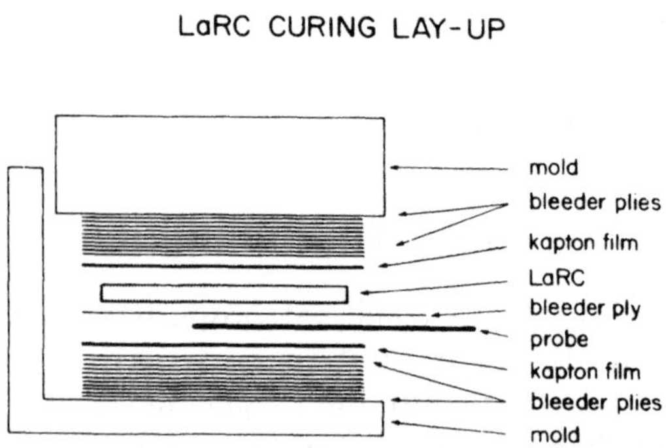

Fig. 6.2. Diagram of a lay-up used inside the mould for measuring the electrical properties of curing polymer systems.

Digital Multimeter. Measurements of capacitance (C_p) and conductance (G_p) at frequencies from 5 to 5×10^6 Hz are taken at regular intervals during the cure cycle and stored on a computer disk. The complex permittivity is calculated for each of these measurements. The temperature is also recorded for each measurement. Plots of the results are prepared from the stored data and are printed using a Hewlett–Packard 7475A six pen plotter. This equipment was used to obtain the experimental data discussed in the latter part of this chapter. Some other equipment suppliers include Gen Rad, IMASS, MICROMET, and Tetrahedron.

6.5 THEORY

The impedance of a resin, as measured by its response to an applied oscillating electric field, will in general arise from two types of molecular processes; ionic and dipolar.

Using either bridge or time-domain techniques, the real and imaginary components of the material's macroscopic impedance Z^* is determined as a function of frequency. Usually, these components are expressed as C_p and G_p. Knowledge of the air-filled replaceable capacitance C_0 of the measurement circuit allows calculation of the complex permittivity ε^*, an intensive property which is independent of the sample's size and shape.

$$\varepsilon' = \frac{C_p}{C_0} \qquad \varepsilon'' = \frac{G_p}{\omega C_0} \tag{5}$$

Both the real and the imaginary components of ε^* have a dipolar and an ionic component.

$$\varepsilon' = \varepsilon'_d + \varepsilon'_i \qquad \varepsilon'' = \varepsilon''_d + \varepsilon''_i \tag{6}$$

The dipolar component arises from rotational diffusion of molecular dipolar moments. In its simplest form, the polar component's frequency dependence is represented by a single relaxation time τ and

$$\varepsilon'_d = \frac{\varepsilon_0 - \varepsilon_\infty}{(1 + i\omega\tau)} + \varepsilon_\infty \qquad \varepsilon''_d = \frac{(\varepsilon_0 - \varepsilon_\infty)i\omega}{(1 + i\omega\tau)} \tag{7}$$

where ε_0 and ε_∞ are the limiting low and high frequency values of ε^*.

The rate of rotational diffusion or re-orientation of the polar moments is characterised by the relaxation time τ. In most systems, a distribution of relaxation times is observed. The distribution is due to a combination of molecular phenomena, including the presence of more than one polar species, the asymmetric shape of the polar group and intra and intermolecular forces. The intermolecular forces determine the extent to which the motion of the dipoles are correlated with each other. In this case ε_d^* is represented by

$$\frac{\varepsilon_d^* - \varepsilon_\infty}{\varepsilon_0 - \varepsilon_\infty} = \frac{1}{(1 + i\omega\tau)}\beta \tag{8}$$

where β is a measure of the distribution in relaxation times.

Ions are often the dominant influence on the complex permittivity of resins and on their changing electrical properties during the cure process. For example, the presence of mobile ions gives rise to localised layers of charge. Since these layers of charge are separated by very small distances, of the order of Å, the corresponding effective capacitance can be extremely large. It is not uncommon to observe values of ε', for example, of the order of 10^6 where values of ε' less than 10 are common for typical liquids and solids. Similarly the translational diffusion of ions significantly increases the conductance, and thereby gives rise to values of ε'' which are equally large, particularly as the frequency decreases.

Most resins, by the fact that they are good insulators, will pick up significant amounts of ionic impurities. Removal of the ionic impurities in a dielectric requires extensive purification steps.

The magnitude of the electrical impedance of the resin is often strongly dependent on the presence of these ions. Many early reports on the electrical characterisation of the cure process neglected the role of ions. This may be due to the extensive solid state literature on widely used polymers (polyethylene, polyesters, polyamides, etc.) which focuses on these polymers' dipolar properties. This was indeed unfortunate. Almost all recent work shows that ionic processes play a major role in determining the impedance during the cure process. This will be discussed in greater detail in Section 6.6, where it will be seen that ionic effects often dominate the electrical impedance, particularly at low frequencies.

The literature discussing attempts to develop equivalent circuit and molecular models to represent ionic phenomena is long and complex.[18–26] Simply summarised, ionic systems involve nonequilibrium and irreversible processes, dependent upon the types, concentration and mobilities of ions as well as on electrode dimensions and discharge conditions. Stated another way, the ionic properties are determined in part by charge diffusion, the hindered translational mobility of the ions in the bulk, and their interaction with neighbouring charged species. Secondly, they are determined by space charge, the build up of charged ionic regions around the electrodes, as well as any other material interfaces and the discharge properties at the material interface. No single or unified representation of all of these effects has been widely accepted. While limited and approximate, one convenient representation of these effects is to consider them as an effective electrode impedance.[19] In this case

$$\varepsilon_i' = C_0 Z_0 \sin\left(\frac{n\pi}{2}\right)\omega^{-(n+1)}\left(\frac{\sigma}{8{\cdot}85 \times 10^{-14}}\right)^2 \tag{9}$$

where C_0 is the replaceable capacitance in Farads, $Z = Z_0(i\omega)^{-n}$ is the electrode impedance induced by the ions and n is between 0 and 1. The imaginary part of the ionic component has the form

$$\varepsilon_i'' = \frac{\sigma}{8{\cdot}85 \times 10^{-14}\omega} - C_0 Z_0 \cos\left(\frac{n\pi}{2}\right)\omega^{-(n+1)}\left(\frac{\sigma}{8{\cdot}85 \times 10^{-14}}\right)^2 \tag{10}$$

σ is the conductivity ($\text{ohm}^{-1}\,\text{cm}^{-1}$), an intensive variable, in contrast to conductance $G(\text{ohm}^{-1})$ which is dependent upon cell and sample size. The first term in eqn (10) is due to the DC conductance of ions translating through the medium. The Z_0 term is due to electrode impedance effects. This term becomes increasingly significant as the frequency of measurement is decreased.

6.6. APPLICATIONS

The remainder of this chapter will focus on three dynamic dielectric measurement parameters which can be used to monitor the cure reaction. In assessing each method it is important to remember that any procedure which is developed and used for cure process monitoring is, in general, unique to a particular resin system and/or application. The essential requirement is that the molecular basis for the frequency dependence of the dielectric signal be understood. Using this information, it will be shown how the ionic contribution ε_i^*,

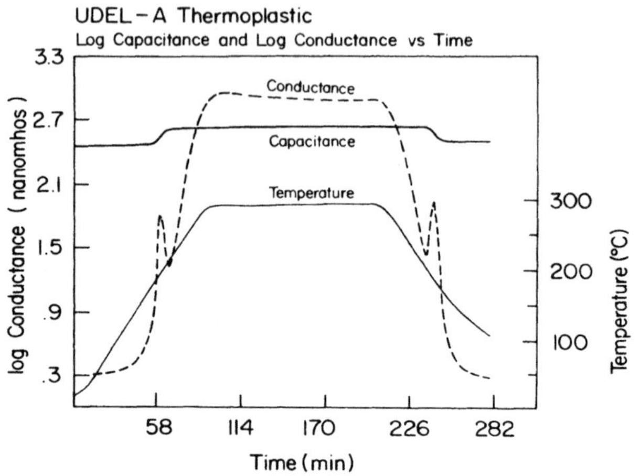

FIG. 6.3. Log capacitance and log conductance vs. time for the thermoplastic UDEL P-1700. The temperature was ramped to 280°C and held for 2 h. A portion of the cool-down curve is also plotted.

the dipolar contribution ε_d^* and the magnitude of ε can be used for cure cycle monitoring.

The first system for which these three dielectric parameters are discussed is the polysulphone thermoplastic UDEL-P1700. Figure 6.3 is a plot of conductance and capacitance at 5 kHz during a 2°C/min ramp to 280°C. The variation in C_p and G_p clearly shows that the magnitudes of both C_p and G_p are a sensitive means of detecting the onset of softening at 150°C and detecting the glass transition at 190°C. The spike in the conductance curve virtually coincides with the temperature of the glass transition. However, care must be taken before using this peak as the glass transition point, without first understanding the ionic and dipolar contributions to the frequency dependence of the dielectric properties in the UDEL polysulphone system. For example the position of the conductance spike varies with frequency.

Figures 6.4 and 6.5 show that plots of ε' and ε'' are complex functions of frequency and temperature. Their molecular basis can be understood if we separate the ionic and dipolar contributions to ε' and ε''. Then the temperature dependence of ε' and ε'' can be used with equations which model the frequency dependence of ε' and ε'' (eqns 6–10) to determine processing properties.

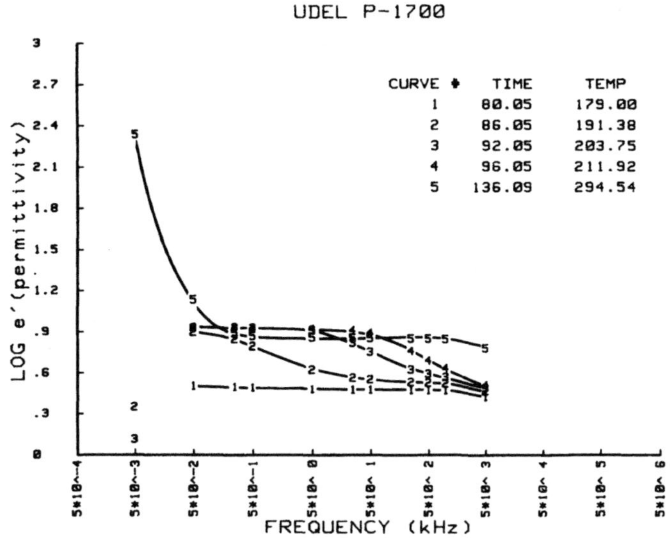

FIG. 6.4. Log ε' vs. frequency (kHz) for UDEL at selected times during a heating ramp to 320°C.

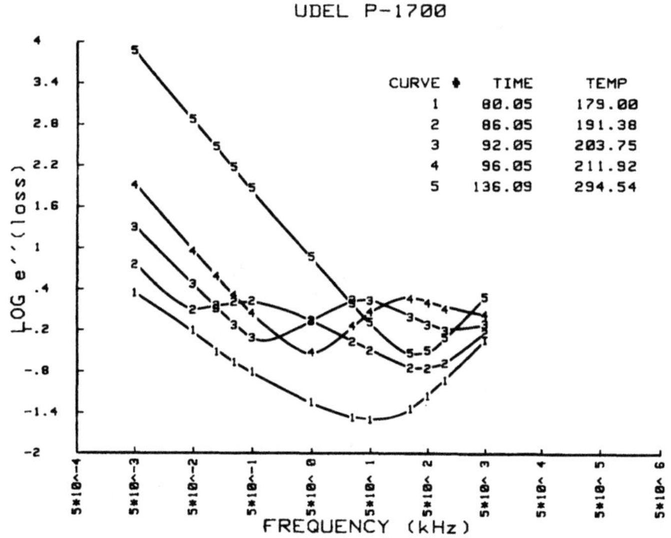

FIG. 6.5. Log ε'' vs. frequency (kHz) for UDEL at times and temperatures of Fig. 6.2.

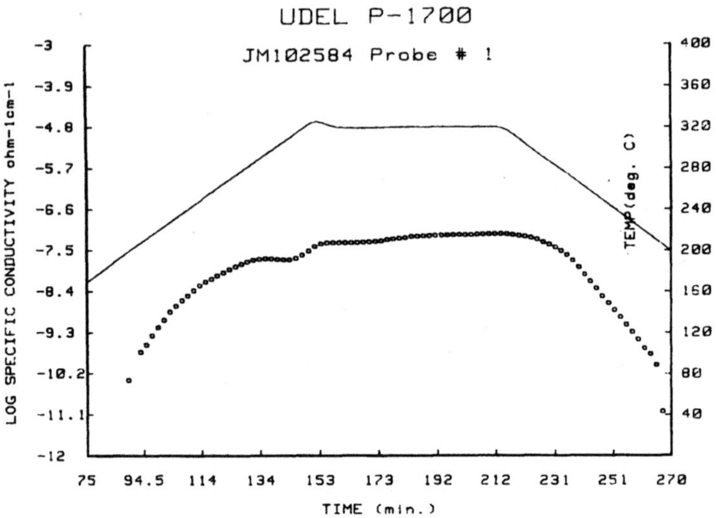

FIG. 6.6. Log specific conductivity vs. time for a temperature ramp to 320°C, a 20 min hold, and cool-down to 200°C.

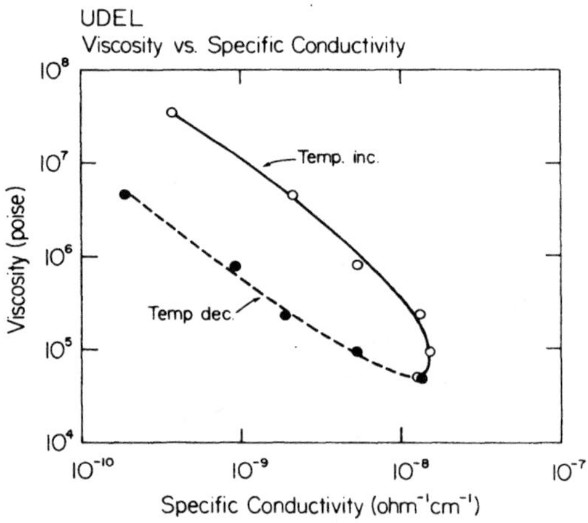

FIG. 6.7. A comparison of viscosity and specific conductivity for UDEL as the polymer is warmed then allowed to cool. The specific conductivity was taken from Fig. 6.4 at points which correspond in temperature to temperatures at which viscosity data was taken. Viscosity was measured on a Rheometrics System 4 Rheometer.

Using eqn (10), and assuming the second term is small compared to the first, the specific conductivity σ has been determined from the frequency dependence of ε''. The specific conductivity σ is a molecular parameter which is independent of sample size. It is a function of the ease with which charged particles translate through the polymer. As such, it is a sensitive measure of the viscosity and the structure of the polymer as the polymer's properties change with temperature (Fig. 6.6). Figure 6.7 is a plot of σ versus viscosity. For polysulphone the product σ_η is a constant at temperatures above T_g. As seen in Figs. 6.6 and 6.7, frequency dependent dielectric measurements may be used to monitor the viscosity quantitatively during processing.

Figure 6.8 is a plot of the dipolar contribution to ε''_d at several temperatures. The plots of ε''_d versus frequency show peaks which move from lower frequencies to higher frequencies as the temperature increases. The reciprocal of the frequency for ε''_d (max) is a characteristic relaxation time τ for rotational diffusion of dipoles. Like σ, τ is a sensitive measure of the viscosity and the molecular structure of the

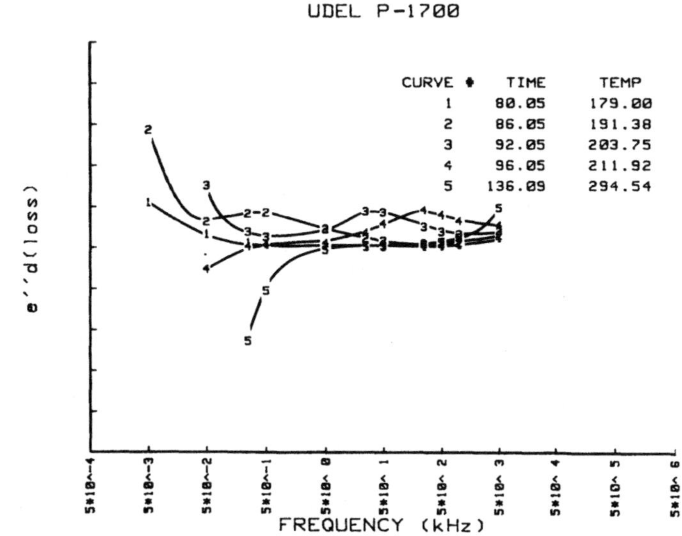

FIG. 6.8. The dipolar loss ε''_d vs. frequency for UDEL at the temperatures of Figs 6.2 and 6.3. ε''_d was calculated neglecting the second term in eqn (6.5). Thus, at the low frequency end of the high temperature curves (curves 4 and 5) a precipitous drop in 'ε''_d' denotes the onset of significant electrode polarisation effects. Units: time, minutes; temp, °C.

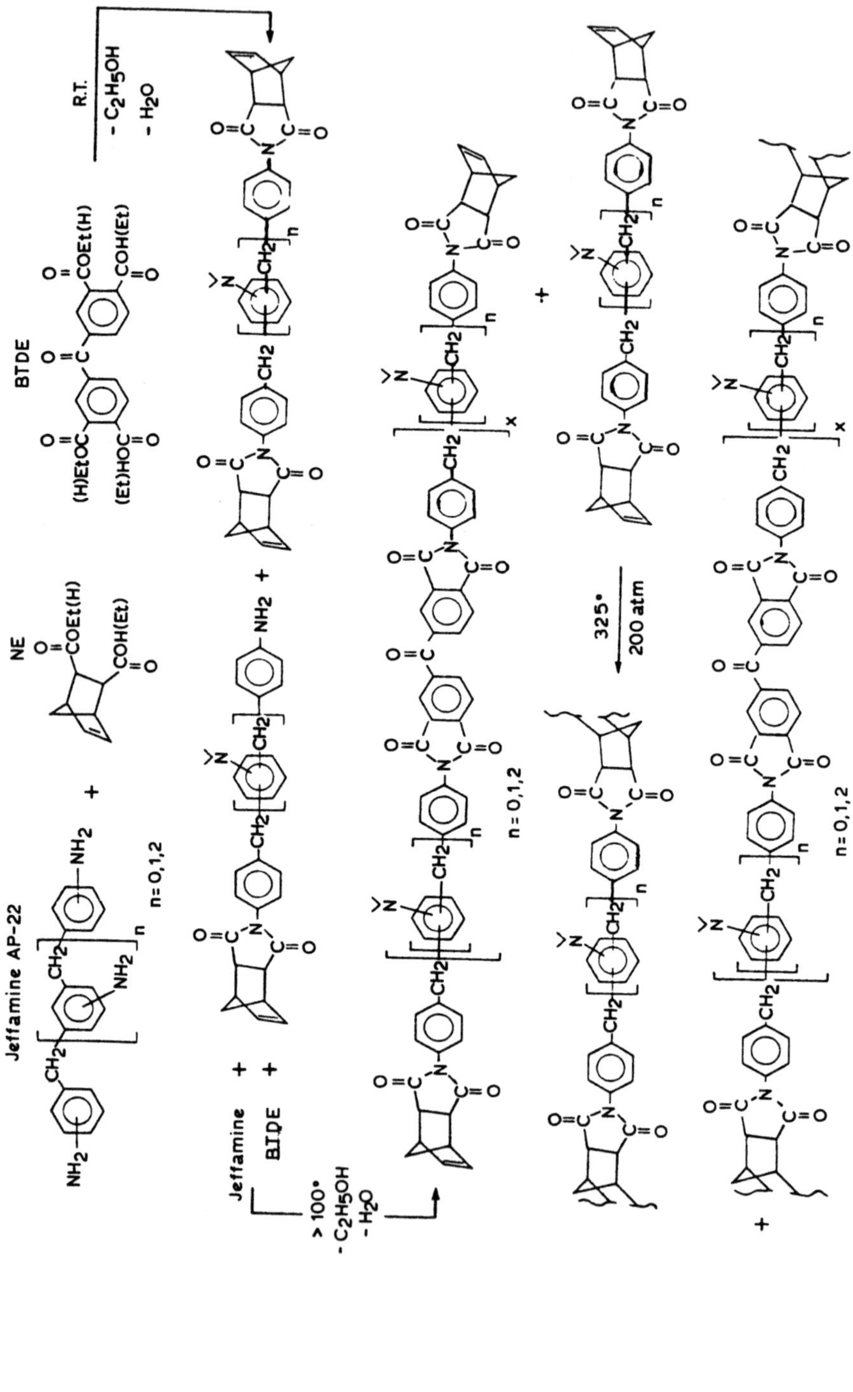

FIG. 6.9. Reaction sequence for polymerising LARC-160.

resin. In turn τ like σ can be used to monitor the state of a thermoplastic polymer system. In Fig. 6.8, one observes that the dipolar relaxation time becomes increasingly short (ε''(max) occurs at successively higher frequencies) as the viscosity decreases and the temperature increases.

The second system examined is the polyimide resin LARC-160. Here the chemistry and flow during the time–temperature cure process is observed. Previous work has shown that LARC-160 polymerisation occurs in 3 steps: first, imidisation of the nadic group and the amine; second, imidisation of the benzophenone tetracarboxylic acid–ester with the amine; third, addition-crosslinking of the nadic end-capped oligomers (see Fig. 6.9). Figure 6.10 plots ε' versus time for several frequencies.

The figure shows an initial rapid rise in ε' due to the onset of flow and the subsequent decrease in viscosity with temperature at 80°C. The rate of increase in ε' with time decreases due to the onset of imidisation of the nadic–amine monomers. At 100°C the rapid rise in ε' resumes, indicating that the nadic imidisation is nearly complete. At 120°C this rapid rise in ε' ends and ε' begins to decrease with time. At this point the onset of imidisation of the benzophenone tetracarboxylic acid–ester and the amine has begun. This chain propagating reaction increases the molecular weight, driving the viscosity up, and ε' down. The point at which the reaction is complete may be monitored by observing the time needed for ε' to achieve a steady state value. From

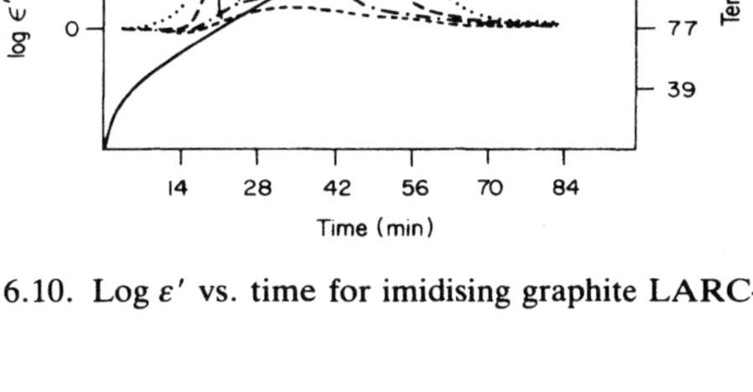

FIG. 6.10. Log ε' vs. time for imidising graphite LARC-160.

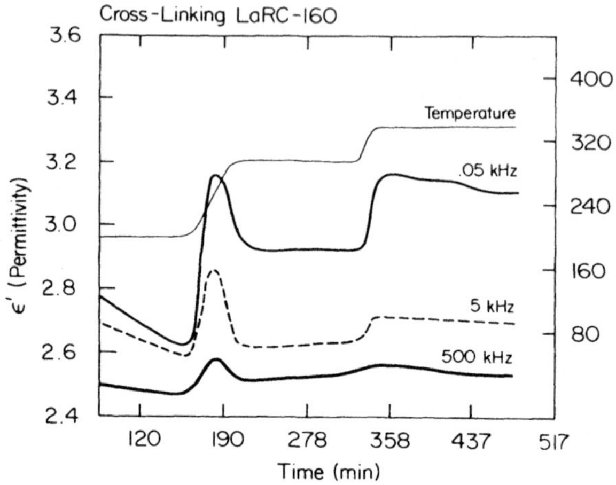

FIG. 6.11. ε' vs. time for curing LARC-160 during 180°C hold and subsequent crosslinking temperature cycle.

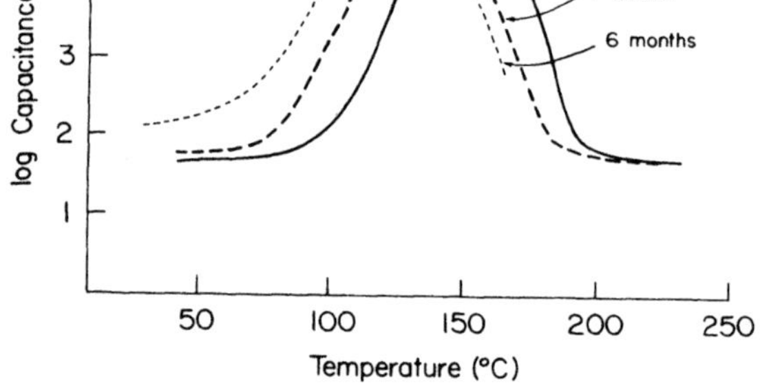

FIG. 6.12. Log ε' vs. temperature for imidising LARC-160 during 2°C/min temperature ramp. The uncured resin was stored at 5°C for times indicated in the drawing.

Fig. 6.11 it can be seen that the reaction at 180°C is still going on at the end of the 1·5 h hold. The high temperature addition reaction can be monitored, as shown in Fig. 6.11. Completion of the reaction during the 280°C hold and the 330°C hold can be determined from the ε' versus time plot. An additional use of the time–temperature dependence is seen in Fig. 6.12. Figure 6.12 shows ε' versus time–temperature for LARC-160 resin stored for varying lengths of time in a refrigerator. Over extended storage periods, the extent of the imidisation reaction, and the associated cure cycle processing properties are advanced in time. These observations, revealed by the dielectric data, are in agreement with other chemical characterisation studies.

These results show that changes in magnitude of dynamic dielectric properties are probes of properties such as the onset of flow, onset of reaction, the reaction completion, as well as the effects of resin age.

The third system to which dielectric analysis is applied is the DGEBA type epoxy resin Epon 828, cured with an amine, Agent U. Figures 6.13 and 6.14 plot ε' and ε'' at three frequencies versus time during a 30°C isothermal cure. The magnitude of ε' and ε'' clearly shows that the reaction is still not complete after 200 min. Figure 6.15 is a plot showing the variation of the modulus G^* and the viscosity

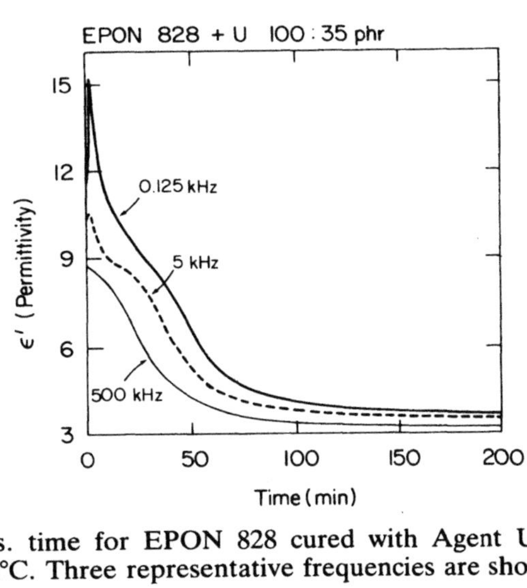

FIG. 6.13. ε' vs. time for EPON 828 cured with Agent U (100:35 phr) at 30°C. Three representative frequencies are shown.

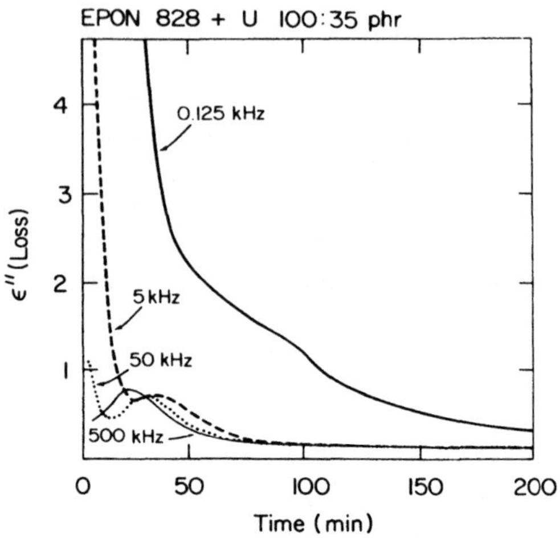

FIG. 6.14. ε″ vs. time for the same sample as in Fig. 6.11. Results for four frequencies are shown.

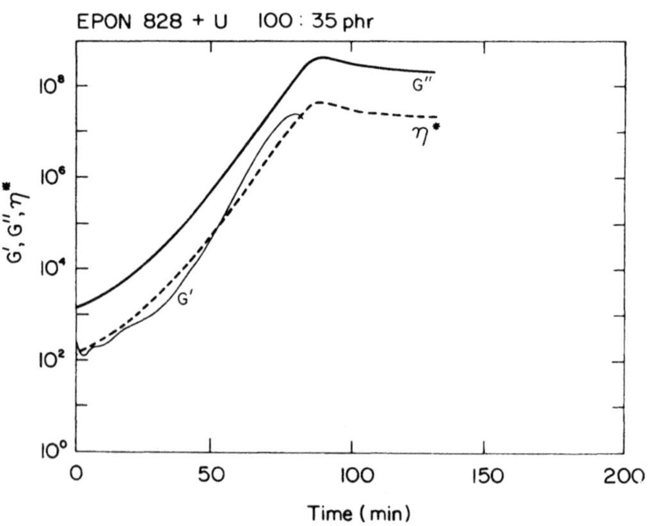

FIG. 6.15. The real and imaginary parts of the modulus and the complex viscosity for EPON + Agent U under the same conditions as above.

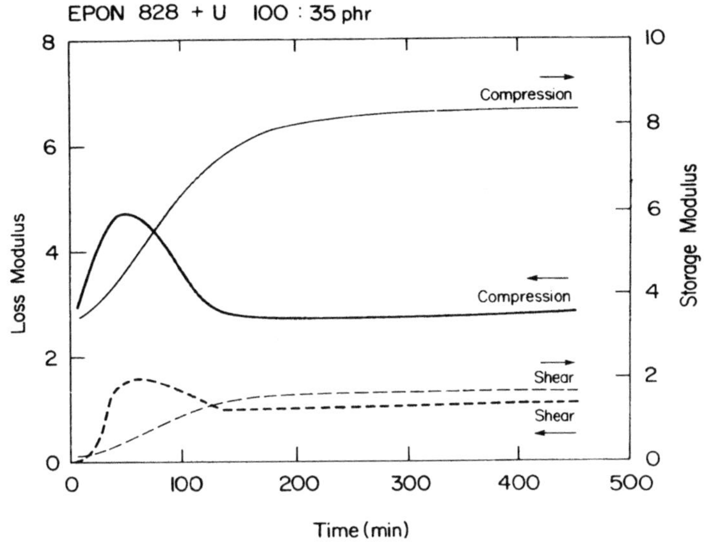

FIG. 6.16. The loss and storage modulus for EPON 828 + Agent U under conditions as above.

over the first 100 min of the 30°C isothermal cure. It is interesting to point out that the modulus and viscosity have moved beyond the measurement range of the parallel plate mode of the rheometer after 100 min. Thus, while the mechanical rheometer in this mode cannot monitor further cure, dielectric techniques can. The fact that further cure is occurring is verified by examining ultrasonic measurements (Fig. 6.16) of the variation of the modulus in both a shear and storage mode over 500 min. Thus the magnitude of ε^* does monitor the cure process continuously over the 400 min needed for complete reaction. During this period the resin goes from a moderately viscous fluid ($\approx 10^2$ poise) to a hard brittle solid ($\geqslant 10^9$ poise).

As discussed for polysulphone, the frequency dependence of ε^* can be used to separate the ionic and dipolar contributions to ε^*. In turn, the molecular parameters σ and τ, which characterise the ionic and dipolar processes for the epoxy resin can be determined. Figure 6.17 plots η versus σ for Epon 828 + U. Figure 6.17 shows that there are two distinct relationships between σ and η. A sharp break occurs when the epoxy gels, often associated with the point $\eta = 10^3$ poise. Up until this point, σ is strongly dependent on η. After gelation σ is now

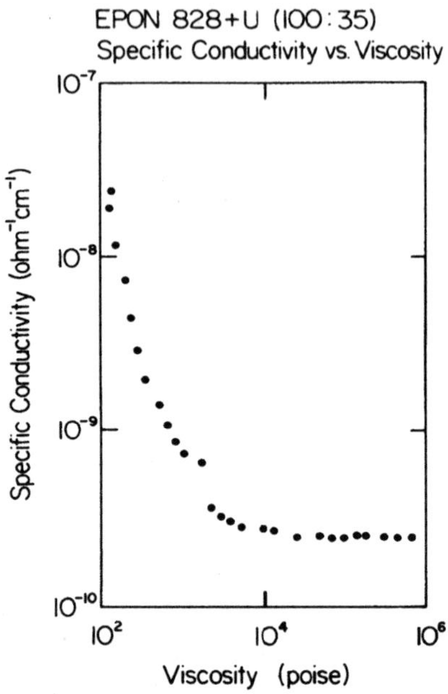

FIG. 6.17. A comparison of specific conductivity from DDA data (Fig. 6.12) with viscosity from rheometrics data (Fig. 6.13) for EPON 828 + Agent U.

sensitive to the effect of increases in crosslink density on the ions' rate of translational diffusion. Thus, its functional dependence on η differs markedly from the pre-gel stage.

In the epoxy resin system, dynamic dielectric analysis can be used to monitor the time–temperature dependence of the polymerisation process. The time it takes for the dipolar relaxation time $\tau = 1/2\pi f_{max}$ to occur can also be used to monitor the extent or stage of the epoxy reaction. Table 6.1 reports the elapsed time for the values of

$$\tau = \frac{1}{2\pi 5 \times 10^5} \quad \text{and} \quad \tau = \frac{1}{2\pi 5 \times 10^3}$$

to occur. A relaxation time

$$\tau = \frac{1}{2\pi 5 \times 10^3}$$

TABLE 6.1
ELAPSED TIME TO ACHIEVE MAXIMUM VALUE OF ε'' FOR DIPOLAR RESONANCE

Maximum in ε'' vs. time at	*Epon 828: Agent U*			
	100:15	*100:24·3*	*100:25*	*100:35*
5 KHz	141 min	89 min	84·9 min	26·1 min
500 KHz	55 min	45 min	43 min	21·7 min
Difference	86 min	44 min	41·9 min	4·4 min

for the resin corresponds to a higher viscosity and a more advanced reaction state than τ(500 kHz). Table 6.1 shows that the elapsed time until the resin attains this relaxation time is an extremely sensitive means for detecting small differences in formulation, and the precise time for a particular extent of reaction to occur.

A final observation on the value of dielectric measurements to polymer processing is shown in Fig. 6.18. The dissipation factor D is the ratio of $\varepsilon''/\varepsilon'$ and as such is an extremely sensitive monitor of physical and chemical changes in the resin and their subsequent effect on ε'' and ε'. As seen in Fig. 6.18, it can be used to detect and monitor the evolution of volatiles. The xylenes vaporise from the poly-

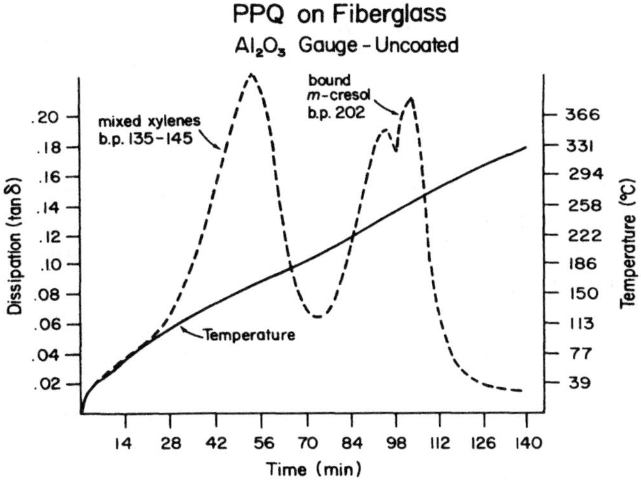

FIG. 6.18. Dissipation vs. time for PPQ film heated with the temperature ramp shown in the drawing.

phenylquinoxaline film at their boiling point. The *m*-cresol, which interacts more strongly with the polymer, is released 50°C above its boiling point.

The use of dynamic dielectric analysis to determine ionic and dipolar molecular parameters, and thereby to monitor the viscosity and extent of reaction in representative thermoplastic and thermoset resins, has been discussed. Several cautionary remarks regarding electrical measurements as a means of monitoring curing resin systems need to be made. First, the use of bulk measurements of C_p or G_p are dependent on variations in sample size and geometry, and are not intensive properties of the material as are ε^*, σ and τ. Second, measurements which rely on a single frequency or a small range of frequencies will in general not correlate with a particular processing property except over a narrow time–temperature range of the cure process. This is because the different molecular processes affecting the ions and dipoles are moving in and out of the time scale of the measurement window. Third, for ionic systems, the ratio $\tan \delta = \varepsilon''/\varepsilon'$ is a poor quantity to initially use to understand and monitor molecular parameters, since it combines the ionic and dipolar phenomena governing ε' and ε'', as seen in eqns (6)–(10). Fourth, measurements at low frequencies (<10 Hz) become increasingly sensitive to electrode impedance effects. Thus, as the frequency becomes lower, a larger portion of ε^* is determined by electrode polarisation rather than by the ionic and dipolar dielectric properties of the bulk material.

6.7. CONCLUSIONS

Recent work has shown that dynamic dielectric equipment can be constructed and used to monitor the cure and fabrication process in a variety of thermoplastics and thermosets, on the pure resin or resin/graphite or glass cloth, over temperatures up to 350°C and covering values of ε^* from 10^{-3} to 10^6. The use of three dielectric parameters which characterise ionic and dipolar molecular processes and thereby monitor the cure process has been discussed. The magnitude of ε^* and two molecular parameters, the ionic conductivity σ and the dipolar relaxation time τ can be used to monitor and measure the viscosity, the state of the system, and the extent of the reaction as a function of time and temperature for representative thermoplastics and representative epoxy and polyimide thermosetting resins. The results suggest that dynamic dielectric analysis provides an extremely powerful and convenient technique for nondestructive quality control monitoring of cure processing.

REFERENCES

1. Kranbuehl, D., Delos, S., Yi, E., Mayer, J., Jarvie, T., Hou, T. and Wintree, B., *SAMPE Symp. Ser.*, **30,** 638 (1985).
2. Kranbuehl, D., Delos, S., Jue, P., Jarvie, J. and Williams, S., *SAMPE Symp. Ser.*, **29,** 1251 (1984).
3. Kranbuehl, D., Delos, S. and Jue, P. K., *SAMPE Journal,* **19** (4), 18–21 (1983) and *SAMPE Symp. Ser.*, **28,** 608–14 (1983).
4. Kranbuehl, D., Delos, S., Jue, P. K., Schellenberg, R. and Mittal, K. (eds), *Polyimides: Synthesis, Characterization and Applications,* Plenum Press, New York, p. 207, 1984.
5. Sheppard, N. F., Coln, M. C. W. and Senturia, S. D., *SAMPE Symp. Ser.,* **29,** 1243–50 (1984).
6. Senturia, S. D., Sheppard, N. F., Lee, H. L. and Marshall, S. B., *SAMPE Journal,* **19** (4), 22 (1983) and *SAMPE Symp. Ser.*, **28,** 851–61 (1983).
7. Sanjana, Z. and Selby, R., *SAMPE Symp., Ser.*, **29,** 1233–42 (1984).
8. Baumgartner, W. and Ricker, T., *SAMPE Journal,* **19** (4), 6–16 (1983).
9. Mitchell, M., *National SAMPE Technical Conference,* **15,** 465–77 (1983).
10. Hedvig, P., *Dielectric Spectroscopy of Polymers,* John Wiley, New York, 1977.
11. Hill, N., Vaughan, W., Price, A. and Davies, M., *Dielectric Properties and Molecular Behavior,* Van Nostrand Reinhold, New York, 1969.
12. McCrum, N., Read, B. and Williams, G., *Anelastic and Dielectric Effects in Polymeric Solids,* John Wiley, London, 1967.
13. North, A., *J. of Polymer Science—Symposium,* **50,** 345–58 (1978).
14. Wada, Y., *Dielectric and Related Molecular Processes Vol. 3,* The Chemical Society, London, pp. 143–75, 1977.
15. Karasz, F., *Dielectric Properties of Polymers,* Plenum Press, New York, 1972.
16. Read, B. E. and Dean, G. D., *The Determination of Dynamic Properties of Polymers and Composites,* John Wiley, New York, 1978.
17. May, C. A., *Chemorheology of Thermosetting Polymers,* American Chemical Society Symposium Ser., p. 227, 1983.
18. MacDonald, J. R., *J. of Chemical Physics,* **54,** 2026–50 (1971).
19. Johnson, J. and Cole, R. H., *J. of the American Chemical Society,* **73,** 4536–40 (1951).
20. Jonscher, K., *J. of Materials Science,* **16,** 2037–60 (1981).

21. JONSCHER, K., *Nature,* **253,** 717–19 (1975).
22. UEMURA, S., *J. of Polymer Science—Physics,* **12,** 1177–88 (1974).
23. POLLAK, M. and GEBALLE, T., *Physics Review,* **122,** 1742–53 (1961).
24. LEWIS, T. J., *Dielectric and Related Molecular Processes Vol. 3,* The Chemical Society, London, pp. 186–219, 1977.
25. NORTH, A., *Chemical and Biological Applications of Relaxation Spectrometry,* Wyn-Jones (ed.), D. Reidel Publishing Co., Dordrecht, Holland, pp. 17–33, 1975.
26. DAY, D., LEWIS, T., LEE, H. and SENTURIA, S., *J. of Adhesion,* **18,** 73 (1985).

Chapter 7

BASIC ASPECTS OF JOINING TECHNOLOGY FOR FIBRE REINFORCED PLASTICS

D. P. BASHFORD

Fulmer Research Laboratories Ltd, Stoke Poges, Berkshire, UK

SUMMARY

In composite constructions there is usually a requirement somewhere for load transfer from one part to another. This is achieved by a means of jointing. In view of the anisotropic and brittle nature of composites this necessitates detailed consideration by the designer.

This chapter identifies the basic parameters in joining composites and the philosophies for efficient design. The majority of developments in jointing techniques have resulted from the uptake of composites for aircraft constructions. Here, the requirements are exacting, and a high level of understanding of composite performance is necessary. Means of jointing can be divided into two groups—adhesive bonding and mechanical fastening—both having their own merits and deficiencies.

7.1. INTRODUCTION

Fibre reinforced plastics can be used in a wide range of structural applications, be they in the aerospace, automotive, ship, building or construction industries. For the transport industries weight saving is the primary reason for using composites, whilst low cost and low maintenance through environmental stability are requirements in building and construction. A galaxy of composites based on carbon, glass and aramid reinforcing fibres are available to meet the specifications of each industry. In conjunction with the fibres different plastics in the form of resins are relevant to each industry. For example, carbon fibres, with aramid and glass supporting, are used with epoxy resins for aerospace applications; these can be considered advanced or high performance composites. By contrast, in ships, buildings and civil constructions, cost constraints dictate that glass fibres and polyester resins (GRP) will predominate.

This chapter will concern itself primarily with advanced composites, notably carbon fibre epoxies (CFRP), and the joining technology evolved by aerospace industries. Most of the general points noted in this chapter are applicable to all continuous fibre composites but, as always, the anisotropic nature of composites should not be forgotten and it must be remembered that both resin and fibre contribute to the overall performance. The joining of GRP is covered in detail by Matthews.[1]

Composites, like other structural materials, must be joined to create useful structures. The function of a joint is to allow load transfer between two components, and ideally, by the most efficient means in relation to either load bearing or stiffness/stability requirements. In conjunction with the anisotropic nature of composites their elastic stress/strain response to failure is significant. This is particularly so for CFRP, where low failure strains of around 1·0–1·5% for both fibre and resin are the norm. Unlike metals there is no plastic deformation which implies that catastrophic failure can occur at ultimate strength. Thankfully, initial failure can invariably be localised (and contained) by the laminate construction used, and in conjunction with damage tolerance requirements, an overall design produced where the risk of catastrophic failure is minimised.

There are two categories of joining applicable to composites: *mechanical fastening,* and *adhesive bonding*. This is true both for composite-to-composite and for composite-to-metal joints, although the joining of different materials (adherends) creates problems, as will be noted. Both categories offer acceptable means of joining composite components, but have different positive and negative features. These features will be outlined as they ultimately determine the suitability and efficiency of a joint design. These latter points can be related to joint weight, load bearing capacity and cost of fabrication.

7.2. JOINING OF COMPOSITES

7.2.1. Two Alternatives

In judging the performance of a joint design, its overall joint efficiency E (the product of its load bearing efficiency, E_L, and its weight efficiency, E_W) can be considered. These efficiencies are defined below:

$$E_L = \frac{\text{load joint can support}}{\text{load continuous member can support}}$$

$$E_W = \frac{\text{weight of continuous member}}{\text{weight of joined member}}$$

$$E = E_L \,.\, E_W$$

Ideally, E_L and E_W should be close to unity. A joint unites two or more members together in order to transfer load between members, and create a useful structure. The joint design can therefore only be evolved in relation to the structure as a whole. The efficiencies above are therefore for guidance and for assessing different designs. The usual means of assessing load bearing efficiencies (E_L) is through test specimens or coupons; this also provides design data, which invariably have to be generated because of the myriad of composite configurations possible.

To choose between mechanical fastening or adhesive bonding the following merits and disadvantages of each type of joint should be borne in mind.

7.2.1.1. Mechanical Joints

On the positive side, mechanical joints:

—permit disassembly without destruction of the members or substrates;
—require no special surface preparation or ultraclean handling operations;
—are not adversely or irreversibly affected strengthwise by thermal cycling and high humidity;
—present no unusual inspection problems for joint quality.

Mechanical joints are less favourable from the following standpoints:

—they require machining of holes in the composite components, which thereby weakens the component;
—they concentrate stress at the bearing surfaces, i.e. where fastener and composite react, thereby creating stress raisers which can initiate failure;
—they are not generally as strong as bonded joints for thin laminates, and are more applicable to thick sections;

—they tend to add weight to the joint, which reduces the overall joint efficiency.

7.2.1.2. Adhesive Bonding

Adhesively bonded joints have the following positive attributes:

—they distribute load over a larger area than mechanical joints, thereby reducing stress raisers;
—they can be designed so as not to weaken the joined members by machining operations in the joint area;
—they add a minimum of weight to the structure;
—after the first loading, bonded joints show less permanent set than equivalent mechanical joints. Good creep resistance requires the correct adhesive selection for the working environment;
—they facilitate the design of smooth external surfaces, and integrally sealed joints with minimum sensitivity to crack propagation;
—on large area joints, bonded assemblies are generally less costly than mechanical joints;
—bonded joints allow for the assembly of dissimilar materials with respect to galvanic corrosion, but thermal stresses due to the dissimilar adherends must be considered.

The disadvantages of adhesive bonding are:

—they are difficult to inspect non-destructively;
—they require designs that minimise peel loadings;
—efficient structural bonds require accurate mating of adherends;
—they are permanent and cannot be disassembled;
—thermal cycling and high humidity can affect their strength;
—they require special surface preparations prior to bonding.

7.2.2. Basic Joint Design

Before designing a joint some basic information is needed. The properties of the member materials (adherends) are required, i.e. the stresses required for failure (in composites) or yield (in metals), the modulus of elasticity, thermal expansion and contraction behaviour, and very often their chemistry if adhesive bonding is being considered. Also, the stresses which the joint is expected to withstand in service and the environment in which the joint is to function must be known. For most structures good joint efficiency in terms of tensile or compressive strength is invariably augmented with requirements for overall structural stability, which may include aspects such as buckling, shear and torsional loadings. On environmental considerations, the coupling of materials with different electrochemical potentials needs to be considered, as this can give rise to galvanic corrosion. This problem exists with CFRP and aluminium, and is not just confined to joining these materials, but also applies to using aluminium fasteners with CFRP. For spacecraft and satellites, performance requirements can be taken a step further, where dimensional stability (constancy of structural dimensions) is required under cyclic thermal regimes (e.g. $-160°C \rightarrow +100°C$) and in a vacuum. Such stability can be achieved with carbon and aramid composites of zero, or near zero coefficients of thermal expansion, provided joints respond likewise. This can eliminate the use of metals with their high positive coefficients of thermal expansion.

When the nature of the materials and the service conditions prevailing are known the selection of the type of joint can be made. This requires a consideration of the limitations imposed by the materials on the processing or manufacturing of the joint. For example, if the joint is to be fabricated in a facility where a manufacturer has no proper surface preparation facility, this would preclude the use of a bonded joint. If, on the other hand, very thin members are to be joined which cannot be machined or weakened with holes in the joint area, then a bonded joint would be in order. Further requirements could include the ability to inspect the joint (e.g. NDE or NDT), strength-to-weight requirements and disassembly capability. The decision whether to use a mechanical or bonded joint requires careful consideration of all variables which will affect the final outcome. This requires a systems approach, with the relationship between the joint and the entire structure kept in mind at all times. Once the selection has been made then the configuration of the joint can be determined. This last process is considered to be the actual design step, but is often embarked upon without proper preparation and guidance.

As a generalisation, wherever possible, adhesive bonding should be used in preference to mechanical fastening. This is primarily due to the damage and weakening that occurs when laminates are cut or machined. Increased susceptibility to interlaminar shear failure results from machining. Along the same lines, preferred design practice for bonded joints requires the surface fibres of a composite adherend to be parallel to the loading direction, to minimise the risk of interlaminar shear failure.

7.3. ADHESIVE BONDING OF COMPOSITES

7.3.1. General Introduction

As an introduction to this section some basic facts on adhesive bonds will be presented. Firstly, the loading modes, or types of stress which an adhesive bond can undergo, are shown in Fig. 7.1. The term 'loading mode' refers to the configuration of the joint and to the way in which members are loaded in relation to each other, whilst 'types of stress' are indicative of what occurs in the joint on a localised basis. Of

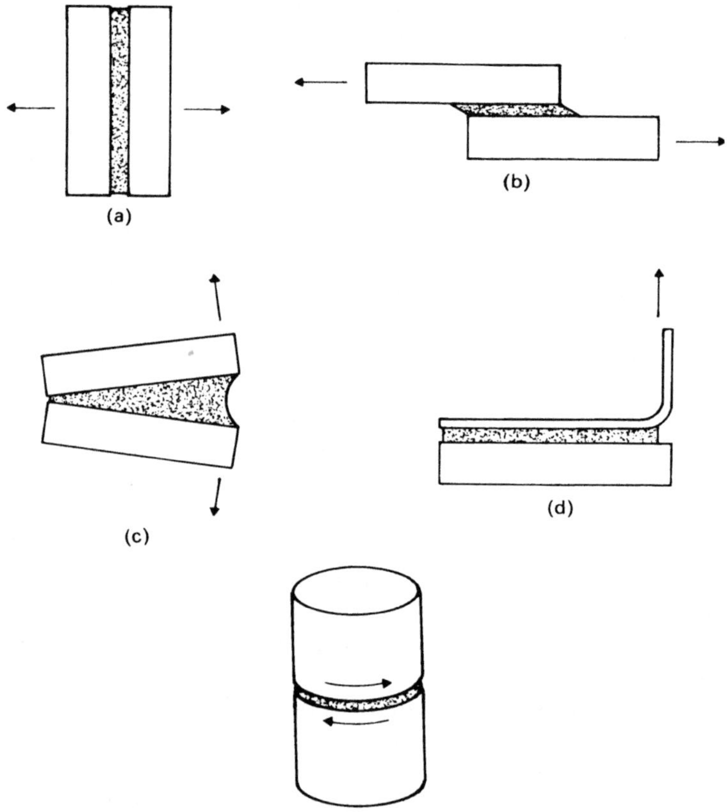

FIG. 7.1. Basic loading modes or types of stress in adhesive joints: (a) tensile; (b) tensile shear; (c) cleavage (thick sections): (d) peel (thin sections); and (e) torsional shear.

these loading modes, tensile shear loading is by far the most desirable for any adherends, but especially for composites. With tensile shear loading, the whole area of the bond can be used to distribute the applied load, and provided the maximum shear stress at any point in the bond is less than that of the failure shear stress of the weakest phase (i.e. laminate resin), the joint will be stronger than the laminate members. However, achieving a pure state of shear in a bonded joint is far from easy, and care has to be taken in avoiding tensile, cleavage and peel stressing somewhere in the joint, if the joint tensile strength is to be maximised and close to that of the base material. Tensile, cleavage and peel loadings all rely on the through-plane tensile strength of a composite laminate, which is a resin dominated property, and hence low. Cleavage and peel loadings also give rise to large stress concentrations in the bond, and the bond area is not fully utilised.

The geometry of the joint and the direction of applied load(s) can cause stress concentrations and introduce undesirable modes of stress, i.e. peel, cleavage and through plane tensile. This reiterates the need

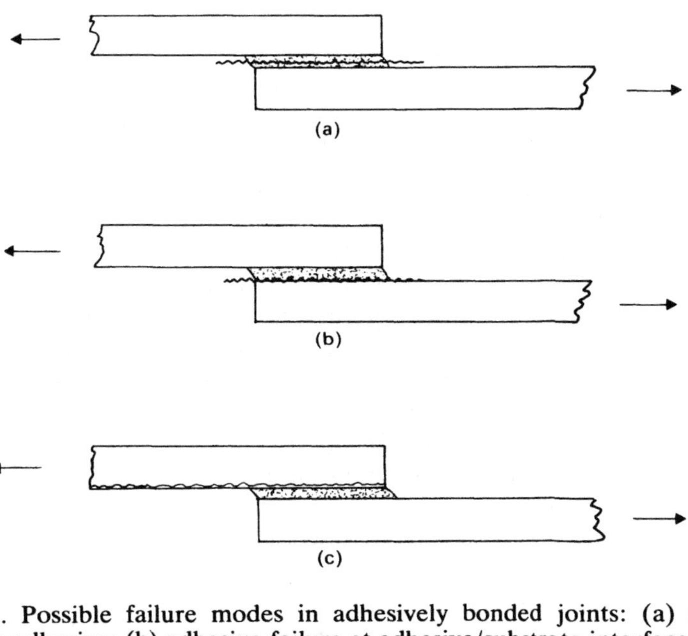

FIG. 7.2. Possible failure modes in adhesively bonded joints: (a) cohesive failure in adhesive; (b) adhesive failure at adhesive/substrate interface; and (c) interlaminal shear.

to consider the joint in relation to the whole structure. The construction of the laminate members in terms of the number of plies and their orientation will affect the stresses within the joint. A large number of joint designs based on the tensile shear mode of loading have been used or theorised for composites. These are outlined in Section 7.3.3 and will be discussed further.

A number of failure modes can occur in bonded composites (see Fig. 7.2). Assuming a tensile failure of the base laminate is being aimed for, these failures could be considered unsatisfactory, but judgement is dependent on the strength required of the joint. Cohesive failure (Fig. 7.2(a)) suggests that the adhesive is functioning incorrectly, possibly as a result of curing problems in the case of epoxy adhesives. Cohesive failure can occur in bonds where peel is occurring, and the adhesive peel strength is exceeded. Adhesive failure (Fig. 7.2(b)) at the adhesive/substrate interface suggests poor surface preparation, provided the adhesive is compatible with the substrate. This is a quality control problem. Interlaminar shear (Fig. 7.2(c)), transverse, or fibre/resin interface failure are more likely modes of failure, and reflect an overstressing of the resin matrix.

7.3.2. Design Aspects for Adhesive Joints

One of the inherent advantages of using composites in structural applications, notably aircraft, is that if the design takes into account the advantageous properties of composites and the manufacturing (moulding) techniques available, the number of individual components in the structure can be reduced compared with an equivalent metallic construction. This has led to the increasing use of single shot mouldings (co-curing) for making fairly complex components to minimise subsequent joining operations, and so reduce costs. These designs can have changes in section, e.g. stiffeners and flanges, which have the appearance of a joint, but rely on direct resin ply-to-ply bonding for their integrity, with no adhesive present. Also, the joint configuration and loading modes prevailing can result in the undesirable peeling and cleavage loading modes. These joints are invariably weak when compared with the potential tensile strength of the laminates used, but are satisfactory when related to the actual working loads.

A large number of publications[2–8] are available which analyse joint configurations and theorise on stress distributions therein. These publications also contain information on the finite element analysis of joints. Because of the complexities of this topic, only the broad findings of joint analyses will be presented, and beneficial features in bonded joint configurations highlighted.

When an adhesive bond is loaded, load transfer is accomplished by minute differential displacements between the adherends. As the adherends are elastic, non-rigid, and distort under load, the displacements across the bond lines are not uniform throughout the bonded area. Hence, the stresses and strains within the adhesive vary over the bonded interfaces. Usually only a small area of adhesive immediately adjacent to the perimeter of the bonded area or at the end of the overlap experiences any significant deformation or accepts significant load transfer. Therefore, bonded joints cannot be designed on the basis of an assumed uniformly stressed adhesive over the entire area. The majority of adhesive in the centre of the bond remains relatively unstressed. This lightly-loaded area should not be looked upon as being inefficient as it is required to resist creep throughout the life of the joint, but increasing the length of overlap to excess will not increase joint strength, as this is largely determined by the stress conditions at the ends of the overlap.

At this point it is worth noting that there are four basic types of bonded joint:

(i) single lap
(ii) double lap
(iii) scarf
(iv) stepped

These joints are shown in their simplest forms in Fig. 7.3, and features specific to each type will be itemised in full later, as a means of summarising the following text.

Work by Hart-Smith[6,8] has been significant in gaining an understanding of how joints perform, and much of what follows is a precis of his findings. The basic sources of non-uniform load transfer within adhesive joints are (1) adherend extensibility in a balanced joint, (2) adherend stiffness imbalance from one end of the joint to the other, and (3) adherend thermal expansion mismatch, as between dissimilar materials. These all need to be considered in conjunction with the elastic trough of lightly-loaded adhesive between the two narrow zones of highly stressed adhesive where the load transfer occurs. These non-uniform adhesive stresses and strains are depicted in Fig. 7.4 for a double strap joint. Sufficient overlap is required for the adhesive shear

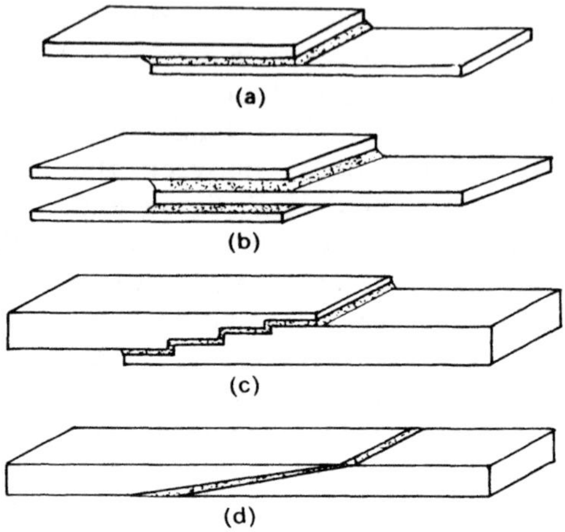

FIG. 7.3. Four basic types of bonded joint: (a) simple single lap; (b) simple double lap; (c) simple stepped lap; and (d) ideal scarf (single taper).

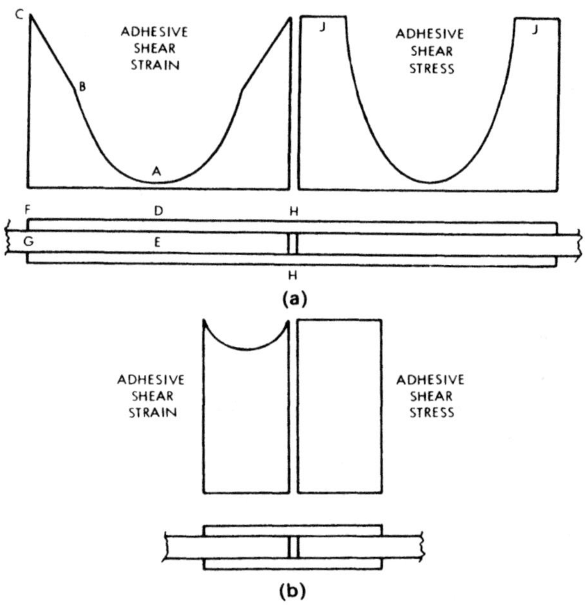

FIG. 7.4. Non-uniform stresses and strains in bonded joints: (a) realistic overlap; and (b) short overlap (reproduced from Ref. 6 by courtesy of Douglas Aircraft Company).

stress in the middle (Fig. 7.4(a)) to be zero or very low. Points A, D and E serve as a memory for the joint, as there is no relative motion here prior to failure elsewhere. In a realistically proportioned joint, negligible relative motion across the adhesive layer at A, D and E occurs, and the relative motion across the bond at C, from F to G, is limited by the differential straining in the adherends. The inner adherend, from G to E, is more highly loaded than the outer adherends, from F to D, and the situation is reversed from A to H. Therefore, the adherend stretches more from E to G than from D to F but, unless the adherends are metal and yield at G or H, the peak bond shear strain is limited by the elastic adherend distortions, no matter how long the load is maintained. Thus creep damage cannot accumulate and the deep elastic trough in the adhesive shear stress distribution is thus vital in ensuring a long service life for bonded structures. When short overlaps are used, the entire adhesive bond strains to the levels at the end of the overlap, ultimately leading to all the adhesive creeping (shearing plastically) and failure occurring. This is what happens in short-overlap coupons where the adhesive is being characterised with the aim of identifying its properties.

The elastic–plastic capabilities of the adhesive should therefore determine the ultimate strength of the joint in relation to adhesive strain energy in shear per unit bond area. This it does, provided the adherends do not fail outside the joint, or peel is introduced, which can fail either adhesive or adherend. Attention now turns to reducing non-uniform load transfer, and hence increasing joint strength. The phenomena discussed in relation to double lap joints are applicable to all joint configurations.

7.3.2.1. Adherend Extensibility in Balanced Joints

Balanced joints have identical adherends of equal thickness and hence stiffness. The extensibility of these adherends affects the adhesive shear strain concentration at the ends of the overlap. Figure 7.5 shows that stiffer adherends promote a more uniformly loaded joint than flexible adherends, in that there is little bond load transfer near the middle of the overlap. The figures are diagrammatical to show the relative displacements and if the overlaps were longer then an elastic trough would be produced. The joint strength is defined by the load at which the adhesive shear strain attains the maximum allowable value, at both ends of the overlap simultaneously.

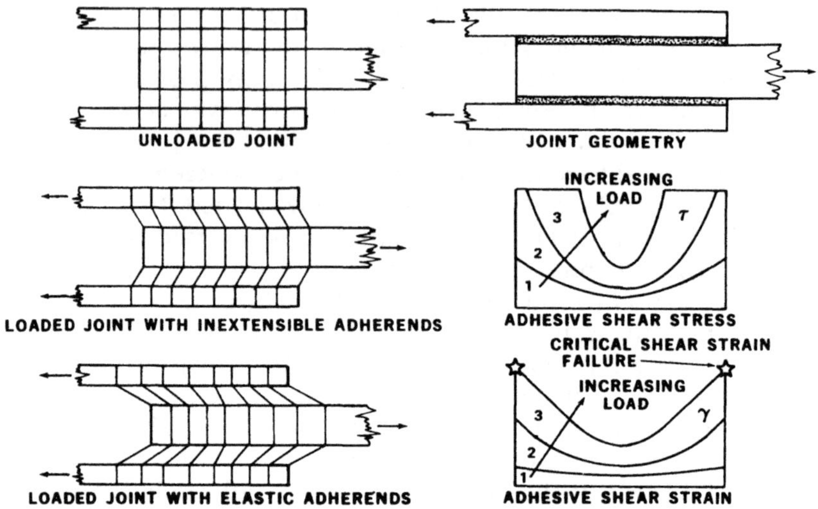

FIG. 7.5. Shearing of adhesive in balanced double lap joints (reproduced from Ref. 6 by courtesy of Douglas Aircraft Company).

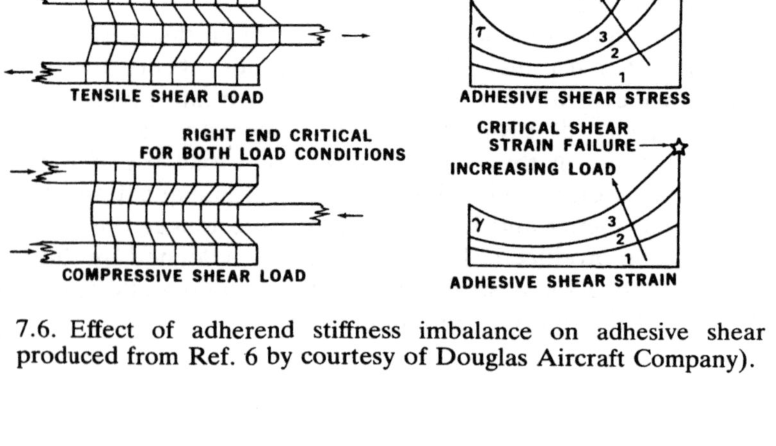

FIG. 7.6. Effect of adherend stiffness imbalance on adhesive shear (reproduced from Ref. 6 by courtesy of Douglas Aircraft Company).

7.3.2.2. Influence of Adherend Stiffness Imbalance

Adherend stiffness imbalance is caused either by joining adherends of the same materials, but of unequal total thickness, or by joining materials of different stiffness. In the former case the total thickness of material entering the joint from either side is important. Figure 7.6 shows how stiffness imbalance intensifies adhesive shear strains at one end of the overlap. The same end of the joint is critical whether the applied load is tensile, compressive or in-plane shear.

7.3.2.3. Effect of Adherend Thermal Mismatch

Adherend thermal mismatch results from joining different materials of different coefficients of thermal expansion. This mainly concerns metal-to-composite combinations. The problems arise in conjunction with high strength adhesives cured at elevated temperatures. Joining effectively occurs at the cure temperature when the adhesive solidifies; on cooling to ambient temperature the differing expansions of the adherends induce residual bond stresses, as shown in Fig. 7.7. The important characteristic of the thermal-stress imbalance is that the critical end of the joint changes with the direction of the applied load.

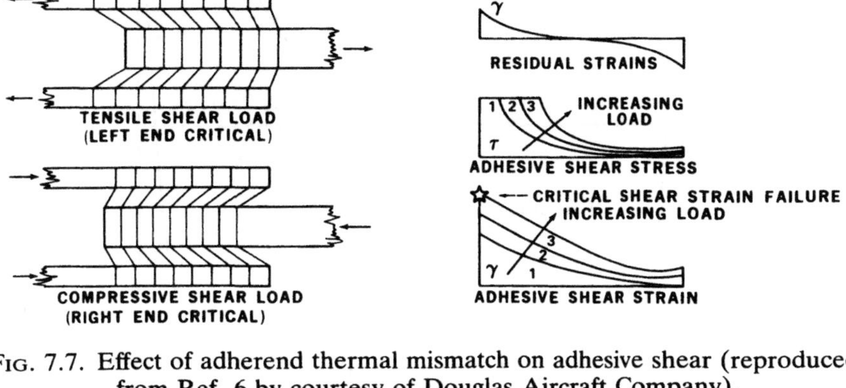

FIG. 7.7. Effect of adherend thermal mismatch on adhesive shear (reproduced from Ref. 6 by courtesy of Douglas Aircraft Company).

If there is an adherend stiffness imbalance and a thermal mismatch then the problems are compounded, especially as the location of the critical strain concentration can change between tensile and compressive loading. For in-plane shear loading, the thermal mismatch effects become most severe in the corners of the bonded area instead of just being an edge effect.

7.3.2.4. Peel Stresses in Bonded Joints

The presence of peel stresses has been highlighted as undesirable. However, they are in fact induced by shear stresses. Peel stresses are

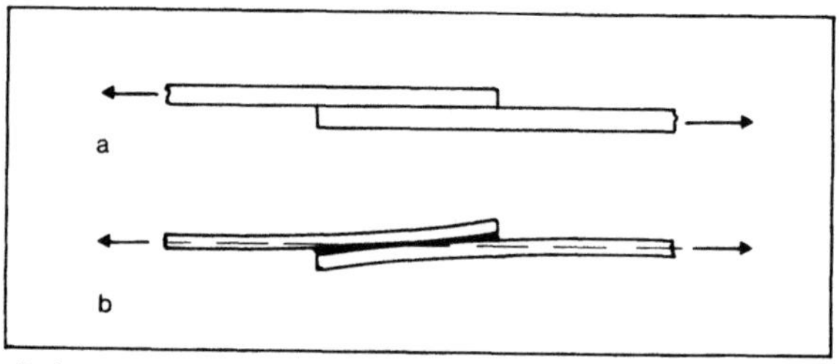

Deformation of a single lap joint: a - loads are initially offset, b - loads ultimately align giving stable configuration

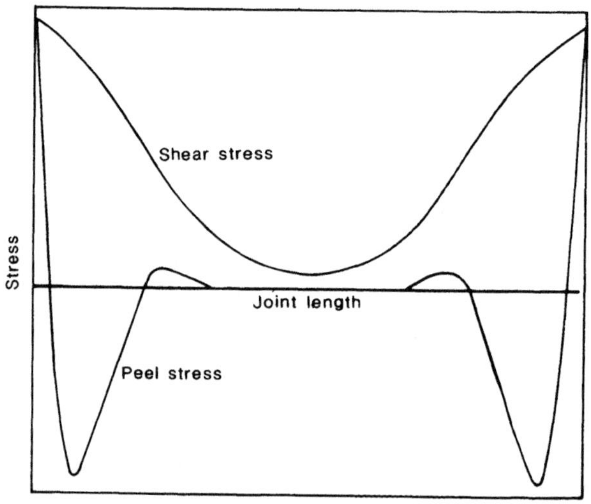

FIG. 7.8. Typical variations of shear and peel stress in a single lap joint (reproduced from Ref. 7 by permission of Butterworth Scientific Ltd).

needed to balance the shift in axis of the shear load in the bond to the stretching load in the adherend. The presence of peel stresses can lead to interlaminar failure in the composite, as it is the weakest phase. With thick sections of metal the adhesive is weaker in peel and fails. The peel stress phenomenon can be approached in a number of ways; (1) by selecting adhesives of high peel strength in the case of metals; (2) by tapering (thinning down) composite adherends at the end of the overlaps; and (3) tapering the end of one adherend away from the other, thus creating a wedge of adhesive which can strain more, and reduce peel stresses (being more compliant). Tapering is particularly effective with single lap and single strap joints where peel stresses are higher because of an eccentric load path.

The response of a single lap joint to loading is shown in Fig. 7.8. Unsupported single lap and single strap bonded joints exhibit a marked dependence on the overlap-to-thickness (l/t) ratio of the adherends. Failure is invariably initiated by the high bending moment induced at the ends of the overlap. For long overlaps the adherends can easily deflect toward the load axis, under a tensile load, to relieve the bending moment. For short overlaps out-of-phase deflections involve higher curvatures and bending stresses, thereby promoting earlier failure. To minimise these bending stresses an l/t ratio of 50 to 100 is typically used. As a rule of thumb, l/t ratios of about 10:1 are associated with joint strengths of only about 30% of the adherend strength, whilst l/t ratios of 50 and 100 give 75% and 90% efficiency, respectively. The high adhesive peel stresses associated with short single overlaps are also decreased significantly for long overlaps.

7.3.3. The Application of Bonded Joints

Analyses of double lap bonds of uniformly thick adherends have established that bonded joint shear strength is proportional to the square root of the adherend thickness, and peel strength is proportional to the quarter power of that thickness, as shown in Fig. 7.9. This shows that thin members, sustaining low loads are limited only by the strength of the adherends. As adherend thickness increases, the margins of excess shear strength and peel strength gradually diminish until there is a peel stress and then a shear stress cut-off. Beyond these cut-offs the ratio of joint strength to adherend strength falls as adherend thickness increases. Ultimately, thick uniform sections should not be bonded, but mechanical fasteners used. To a point the joint can be redesigned to include the more complex stepped lap or

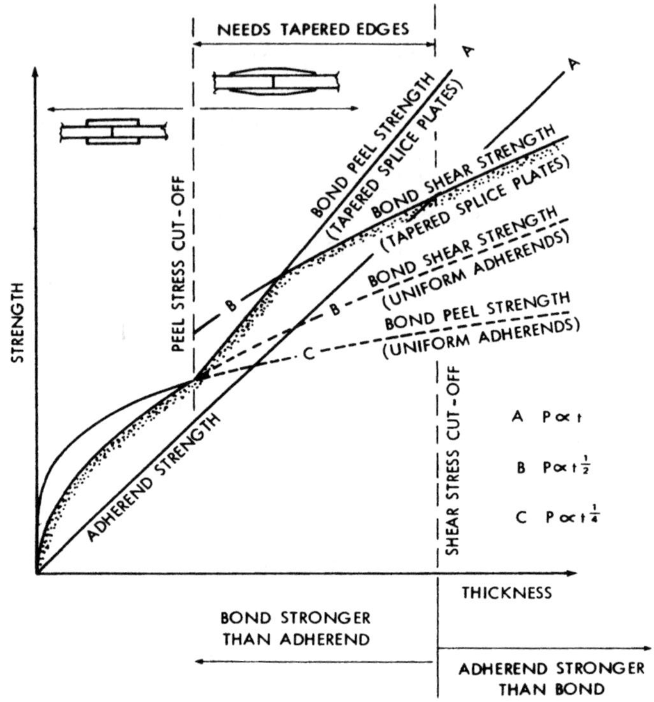

FIG. 7.9. Effect of adherend thickness on bond strength (reproduced from Ref. 6 by courtesy of Douglas Aircraft Company).

scarf configurations to increase load-bearing capabilities. Figure 7.9 is indicative of double lap performance for all adherends, but composites complicate matters by their interlaminar weakness. The use of double lap joints can be fitted into the overall picture of the relative uses of different bonded joints.

7.3.3.1. Bonded Single Lap Joints

These are the simplest of bonded joint configurations, but are only used on the thinnest of laminates. Despite the use of beveling or radiusing, bending of the adherends due to the eccentric load path ensures low joint strength in relation to adherend strength for thicker laminates as a result of interlaminar failure. Figure 7.10 shows variations of the single lap principle which could be used.

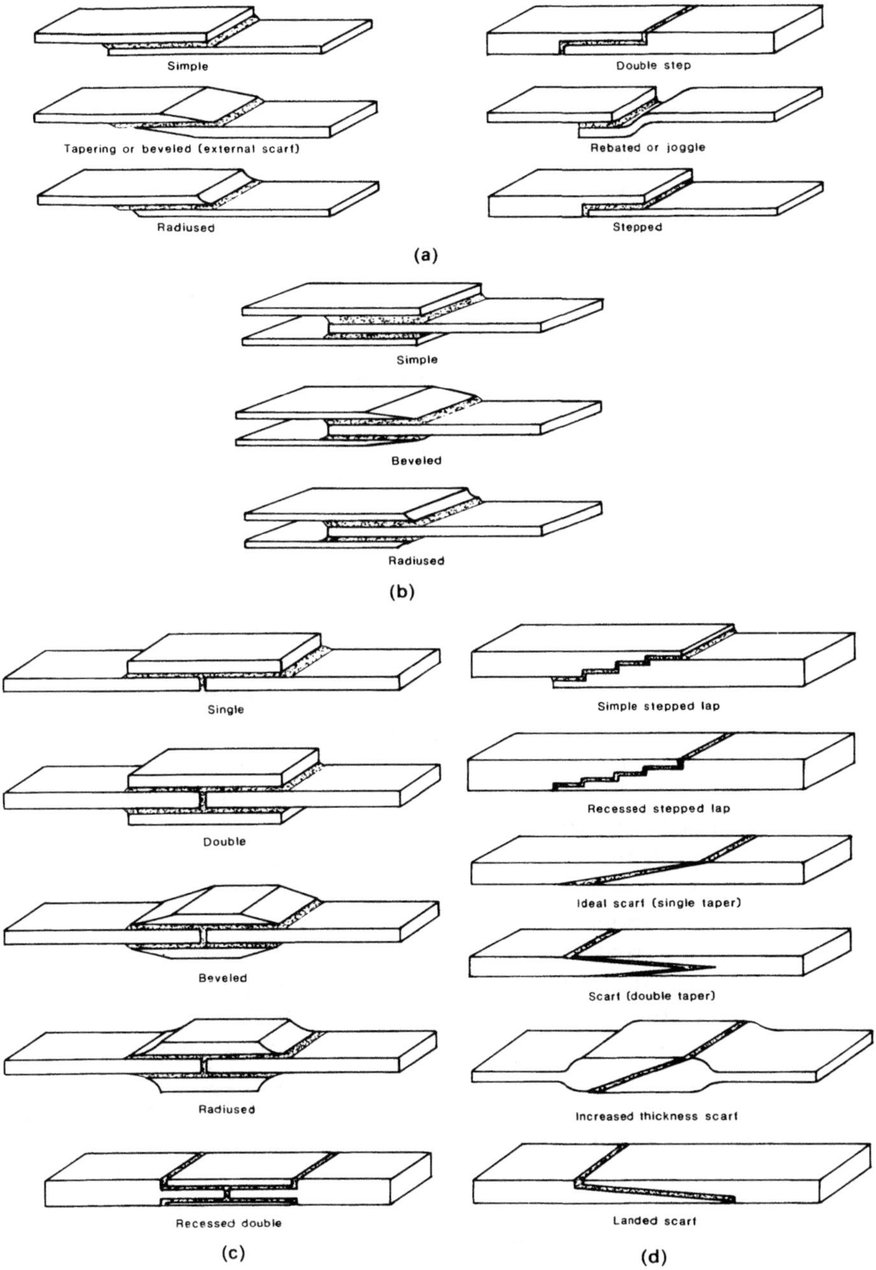

FIG. 7.10. Bonded joint designs: (a) single lap joints; (b) double lap joints; (c) strapped joints; and (d) stepped lap and scarf joints.

7.3.3.2. Bonded Double Lap or Double Strap Joints
In balancing the joint, the full strength of thicker laminates can be utilised. When peel failures limit strength then tapering or bevelling can boost strength as adherend thickness increases until shear failure reduces strength efficiencies. The use of tapering or beveling will increase manufacturing costs, and this has to be balanced against strength requirements.

7.3.3.3. Stepped Lap and Scarf Joints
Stepped lap joints can be considered the next link in the quest for strength efficiency. For these joints the ply stacking sequence becomes important as steps are created through the laminate. For good load transfer the step flats need to coincide with 0° plies assuming this to be the loading direction. Better strain compatibility can be achieved with this gradual change in adherend thickness throughout the bond. Such joints are weight efficient in minimising the amount of joint material; the same can be said for scarf joints.

In theory, a bonded scarf joint approaches the ideal in terms of strain compatibility between the adherends and reducing the stress concentrations at the ends of the overlap. However, in practice it can be difficult to realise the full potential of a scarf joint, because of the practicalities of machining scarf angles, handling laminates and applying uniform pressure to the angled bondline to control bondline thickness. Scarf angles of less than 5° are usually required for the joint strength to match adherend strength. Thin laminates, of the order of a few millimetres in thickness, tend to be precluded from scarfing. The orthotropic nature of laminates is a hindrance in these types of joints, and ply stacking sequence is again significant. The reason for this is that scarf angles create areas of non 0° orientated plies on the faying surfaces. The shear strength of non 0° orientated plies is poor, so that on loading the joint, these areas are susceptible to interlaminar shear failure. Ultimately, for thick sections, mechanical fastening is preferred in giving the required strength at less cost. Scarf joints have been successfully used in composite repairs where a patch is used on thick sections to replace localised areas of damage. High strength stepped and scarf joints produce more complex and exacting inspection requirements because of their configuration and the criticality of structures they can be used in.

7.3.4. Adhesives for Composite Joints

The selection of an adhesive for a specific joint is dependent on the required mechanical properties in relation to the operating environment and compatibility with the adherends.

The mechanical properties of concern are tensile and shear strength and moduli, although in many instances a peel strength requirement will also exist. Analyses of joints has highlighted the use of adhesives with low tensile and shear moduli as being beneficial to joint strength. There is, however, a limit on minimising adhesive modulus because of the possibility of creep and reduced environmental stability, especially in moist conditions, if the modulus is taken too low. It is, however, good practice to use an adhesive with a shear strength some 50% above that of the composite adherend.

For the stress analysis of joint designs the properties of the adhesives considered need to be known. These properties can be obtained from the following tests.[2]

Test method	*Data yielded*
1. Napkin torsion ring test	Ultimate shear strength and shear moduli
2. Thick adherend lap shear	Ultimate shear strength and shear moduli
3. Circular butt joint	Ultimate tensile strength and tensile modulus

For high performance composites, notably CFRP used in aircraft construction, a thin, controlled bondline of approximately 150–200 μm is preferable. Excessively thin or thick bonds tend to be detrimental to strength, although thickened bonds can in some circumstances be used for stress relief at the beginning of an overlap.

On compatibility with adherends, the majority of adhesives used are epoxy based, reflecting the dominant use of epoxy resin composites in high performance applications. Unless polyimide or thermoplastic resin composites are to be joined, then epoxies are always applicable, and this goes for GRP (polyester) although modified acrylics are also used.[9] Epoxies will also readily bond to metals such as aluminium and titanium, but surface preparation is all important for durable and reliable bonds. For composites the best preparation is thorough mechanical abrasion to remove the resinous surface layer down to the first layer of fibres,[2,10–12] followed by solvent cleaning to remove debris. This provides a contaminant free and an adhesive receptive surface. Mechanical abrasion or grit blasting is not always a practical proposition for large or delicate structures. In these cases a nylon peel ply is placed in contact with the uncured preform prior to moulding. In subsequent bonding operations the nylon ply is peeled from the moulding exposing a roughened resinous surface layer suitable for adhesive bonding. Metals invariably require a multi-stage operation involving degreasing, washing with water and some form of acid or proprietary etch.

Structural adhesives are available in two basic forms, as semi-solid ('B' stage) films (some with carrier fabrics) or as thixotropic pastes (two-component systems). The 'B' staged films are cured at elevated temperatures and their elevated temperature performance needs to be compatible with the operating temperatures of the composite adherend. Good control of bondline thickness and quality can be achieved with film adhesives. The paste adhesives cure at room temperature with some possessing post-curing capabilities to 100°C. These adhesives tend to be used where the application of pressure on the bond is impracticable, or the mating of joining parts is uncontrolled.

7.4. MECHANICALLY FASTENED JOINTS

7.4.1. Failure Modes

Many of the early publications on mechanical fastening of composites presented empirical work aimed at maximising joint strengths.[13–15] As CFRP becomes increasingly used in primary aircraft structures, and mechanical fastening is necessary to achieve high joint strengths, the need for theoretical predictions of joint strength[16–20] has increased to avoid protracted and costly testing. In this chapter beneficial features in joints will be identified. It is not possible to characterise all aspects of joint behaviour, but the facts presented allow inferences to be made for types of joint and composites.

Mechanically fastened joints can display a number of failure modes, as shown in Fig. 7.11. These failure modes are invariably gradual, with progressive damage eventually resulting in complete loss of load bearing capacity. The definition of joint strength can become confused, the failure criteria can be the onset of damage, degrees of damage or ultimately the maximum load bearing capacity. In estab-

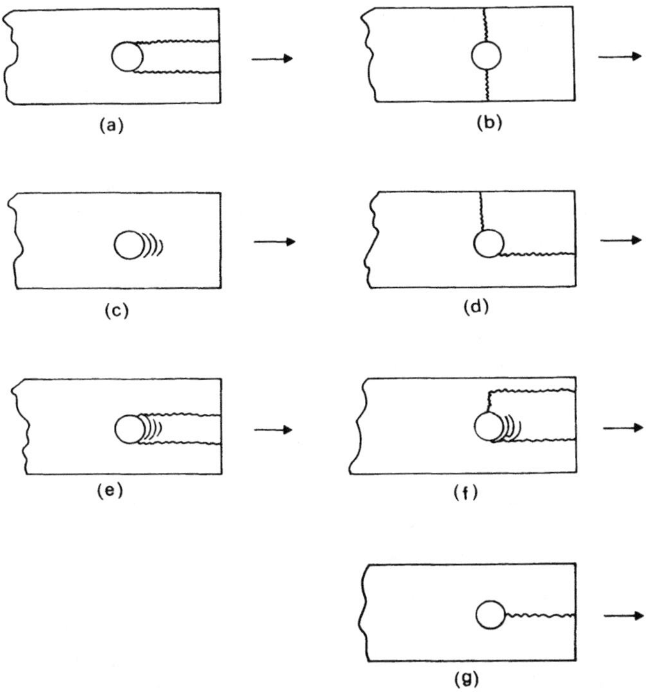

FIG. 7.11. Bolted joint failure modes: (a) shearout; (b) tension; (c) bearing; (d) tension/shearout; (e) bearing/shearout; (f) bearing/tension/shearout; and (g) cleavage.

lishing design data the use of simplified coupon tests is the predominant source. The occurrence of a particular failure mode is predominantly the result of joint geometry.[13] Net-section tension failures occur when the bolt diameter is a sufficiently large fraction of the strip width. This fraction is about one-quarter or more for near-isotropic patterns in carbon-epoxy, but may rise to a minimum of a half or more for glass epoxies. Bearing failures occur predominantly when the bolt diameter is a small fraction of the specimen width. Shearout failures can be regarded as a special case of bearing failures. Frequently a shearout failure results from a bearing failure with a short edge distance. Cleavage failures are another form of low-strength failure, because the fastener is too close to the end of the specimen. It is not uncommon for a cleavage failure to be triggered by an incomplete net-section tension failure. A cleavage failure is invariable initiated at the end of the specimen. Both shearout and cleavage failures are unacceptable, and easily avoided. Instances of bolt heads being pulled through the laminates can occur if the bolts bend. This usually occurs with countersunk fasteners and indicates that the ratio of laminate thickness to bolt shank diameter is too large.

There are a large number of variables affecting the strength of mechanically fastened composites.[2,11,13,14] They can be divided into three groups: material, fastener and design parameters.

7.4.2. Material Parameters

Although many of the general observations made on mechanically fastened joints are applicable to all composites, in that these are anisotropic materials, there are specific differences between carbon, glass and aramid composites. Firstly, aramid composites and high strength mechanically fastened joints do not go together as the inherent compressive strength of aramid composites is low and this inhibits load transfer in bearing. Aramid constructions tend to be thin, tensile stress orientated, and adhesively bonded. Carbon fibre composites are considerably stiffer than glass fibre composites, and hence more severe stress concentrations occur in CFRP than in GFRP around fastener holes. This has important consequences for the initiation of failure. Failure is initiated by a combination of fibre breakage, resin cracking and delamination. For glass/epoxy composites, greater degrees of delaminations coupled with lower modulus and greater fibre strain to failure provide considerable stress relief. For these reasons the use of glass fibre plies instead of some carbon fibre plies in CFRP construction has proved effective in stress concentration relief. Primarily used in the 0° direction the GFRP acts as a softening strip, with uses in bolting aircraft wing skins to spar caps. The response of hybrid composites effectively forms another materials parameter.

In that failure of fastened joints is dependent on fibre and matrix, the fibre form and hence resin content can be significant. Fibre forms are primarily unidirectional plies (tape or prepreg) and woven fabric (cloth). A fabric composite will have a lower volume fraction of fibre, and hence reduced section strength. This, coupled with the various types of weave, weft and warp bias, makes strength prediction much more difficult, and makes coupon testing necessary to acquire design data. The majority of analysis has been conducted on unidirectional ply composites, where orientation and stacking sequence prove to be very significant.

Clearly the 0° direction will be dominant in terms of providing strength, but other orientations are needed in structures, primarily the ±45° directions for shear and torsional support, and 90° for transverse and bearing strength. In addition to the overall structural requirements usually needing three fibre orientations, the same can be said of mechanical joints. It is now well established that ±45° plies reduce stress concentration factors. These plies are also damage tolerant, which is another reason for their introduction into aircraft wing skins, for example. For optimum bearing properties in CFRP >55% but <80% of 0° plies are necessary, the balance being made up with ±45° plies to give transverse stability.[7,21,22] Optimum tensile properties are obtained when the ratio of 0° to 45° plies is 2:1, whilst optimum shear strengths need a ratio of 1:1. Overall, 50–60% of 0° plies are recommended. The introduction of 90° plies complicates matters, but can be necessary to eliminate shearout failure. The combination of 0°, ±45°, 90° is an improvement in terms of the ratio of joint strength to laminate strength. It is also preferable to have the 90° plies on the laminate surfaces, i.e. $(90, \pm 45, 0)_S$.

7.4.3. Fastener Parameters

The choice of fasteners is between bolts, rivets and screws. Bolts are preferred, with screws the least favoured because of the extensive damage done to the laminate. By all accounts screws should never be used, although on cost grounds they could be used with GRP when strength and environmental durability are less critical. Riveted joints give adequate strength in high-duty CFRP laminates in thicknesses up to about 3 mm.[15] The type of rivet, the use of washers, dimensions of countersink and imposed clamping loads all affect the ultimate strength of the joint. Adhesive bonding is preferable to riveting in terms of both strength and cost of manufacture. Bolts offer the greatest strength for mechanically fastened joints, and are preferred for thick sections over any other form of joining. Unless overtightened, no damage is done to the composite during assembly. In terms of optimising strength, the following are beneficial:

—The bolt should be a good fit into the hole.
—Increasing the bolt tightening torque, and hence clamping force, increases the bearing strength. Optimum levels of clamping exist.

—The hole in the washer should be a close fit to the bolt. Washer size can influence strength.

On the point concerning a close fit between the hole in the composite and the bolt, the quality of the machined hole is also important. Because composites are brittle and prone to delaminating special precautions and drill bits are required to create a clean cut.[11] Good alignment and accuracy in machining lines of bolt holes are required to ensure even stress distribution on loading the structure. Stiff, low strain to failure CFRP is unforgiving to localised over stressing in bolted joints.

Whilst increasing the bolt tightening torque raises bearing strength by generating frictional forces between the joined substrates, some relaxation can occur with time. This is primarily caused by resin creep, particularly if the resin softens through moisture absorption. It is therefore prudent to consider the extra load bearing capacity generated through bolt tightening as an unaccounted safety margin during design, unless the strength of the clamped joint has been measured or predicated for the whole life of the structure.

7.4.4. Design Parameters

The choice of joint type, of which a few are shown in Fig. 7.12, is dependent on the particular application envisaged. When using experimental data care should be taken to ensure close correlation between idealised test joints and the actual configuration being designed. All loading modes and eccentricities should be considered. This usually necessitates considerable testing of bolted specimens to produce design data.

Geometric effects in terms of fastener location and spacings have been extensively studied. Figure 7.13 shows the nomenclature used and some of the significant features are noted below for CFRP to CFRP joints:

—End distance effects. For each CFRP laminate construction there is a minimum value of the e/d ratio needed to develop full bearing strength. For pseudo-isotropic lay-ups (with 0°, ±45° and 90° orientations) the ratio is always greater than 3. For just ±45° lay-ups the ratio is greater than 5.

—Width effects. From work on single hole specimens, 0°, ±45° lay-ups in CFRP require a minimum W/d of 4 (W/d of 8 for all ±45° constructions). These measured ratios are assumed by many

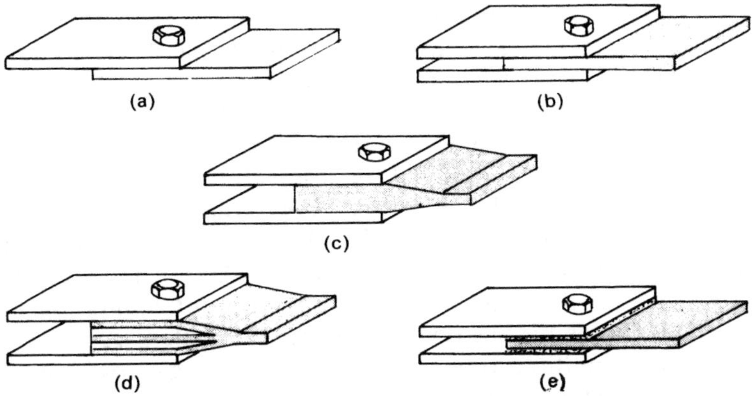

FIG. 7.12. Bolted joint designs: (a) single lap; (b) double lap (including bushings); (c) reinforced edge; (d) shim joint; and (e) bolted–bonded.

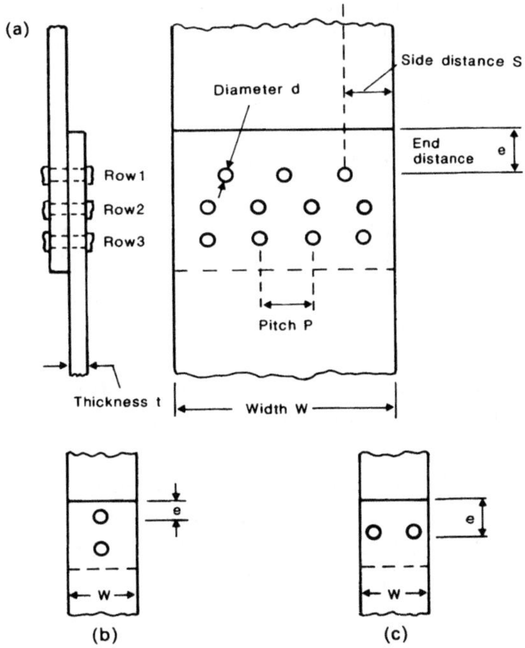

FIG. 7.13. Definition of joint geometry: (a) rows of bolts (rows 1 and 2 are staggered, rows 2 and 3 are in uniform rectangular pattern); (b) two bolts in tandem (line of bolts); (c) two bolts in parallel (row of bolts). (Reproduced from Ref. 14 by permission of Butterworth Scientific Ltd.)

people to equate to minimum pitch distances for multihole joints. In reality they are close approximations. It should be noted that in 0°, ±45° lay-ups the minimum *e/d* and *W/d* ratios will depend on the relative proportions and sequence of 0° and ±45° material.

—As the *d/t* ratio increases, bearing strength reduces. The *d/t* ratio should not exceed unity if full bearing strength is to be developed.

—On hole patterns, the evidence shows that a tandem row (two rows) is preferable to a parallel (Fig. 7.13) and that the use of staggered pitch with tandem rows helps. This is somewhat subjective as the gains are small in terms of optimising bearing strength. Indeed it has been stated that as little is gained by using tandem or multiple rows a well designed single row joint makes more economic sense in terms of construction.

Full bearing strength, for bolted joints, is only developed if failure due to tension and shear (Fig. 7.11) is suppressed by providing adequate end distance and width (or pitch), and if sufficient restraint is present to prevent local interlaminar failure. Simple bolted joints rarely achieve more than 50% of the base laminate strength. The quest for high joint strength can be helped by localised reinforcement and load distribution exercises ((c), (d) and (e) in Fig. 7.12), but these reduce joint efficiency in terms of weight and increase production costs.

The definition of failure varies widely which, in turn, confuses the value of 'design allowable' strengths. A partially accepted failure criterion is based on the deformation of the bolt hole under load. However, agreement as to the value of hole deformation is lacking, with values varying from 0·5% to 4% of the hole diameter being suggested. A safety factor of two on the ultimate strength would appear to be a minimum, and fall close to the stress at which some damage is initiated.

Papers covering stress analysis and modelling of bolted joints are increasing in numbers[16–20] as the prediction of joint performance is preferable to extensive and expensive coupon testing for generating design data, although some testing will always remain to confirm the theory.

7.5. CONCLUSIONS

Fibre reinforced plastics are increasingly used in structures. For all mobile applications efficient means of joining composites are required to utilise fully their weight saving characteristics.[23] The anisotropic nature of composites necessitates careful joint design to avoid interlaminar or shear failures. This requires the designer to consider the joint in respect of the whole structure and the loading conditions imposed on the joint.

The efficiency of a joint can be assessed in relation to its strength, weight and cost of manufacture. A compromise is invariably required, as maximising strength usually leads to increased weight and/or increased fabrication costs. A joint's strength needs to be that required to meet the loads involved.

Both adhesive bonding and mechanical fastening offer acceptable means of joining composite to composite and composite to metal. For thin sections (laminates) less than 3 mm in thickness adhesive bonding is the preferred means of joining. As section thickness and load carrying requirements increase, then bolted joints become more appropriate.

REFERENCES

1. MATTHEWS, F. L., Problems in the joining of GRP, *Developments in GRP Technology—1* (B. Harris, ed.), Applied Science Publishers, London, 1983.
2. DEVINE, A., Joining of advanced composites, *Engineering Design Handbook,* US Army Material Development and Readiness Command, DARCOM PAMPHLET 706-316, 1979.
3. SAGE, G. N., Some aspects of bonded joint design in CFRP, *Composites,* **3,** 256–60 (October 1976).
4. WRIGHT, M. D., Stress distribution in carbon fibre reinforced plastic joints, *Composites,* **11,** 46–50 (January 1980).
5. ALLRED, R. E. and GUESS, T. R., Efficiency of double-lapped composite joints in bending, *Composites,* **9,** 112–18 (April 1978).
6. HART-SMITH, L. J., Adhesive-bonded joints for composites—phenomenological considerations, *Advanced Composites Technology: Proceedings of the Conference,* El Segundo, California, pp. 163–80, 1978.
7. MATTHEWS, F. L., KILTY, P. F. and GODWIN, E. W., A review of the strength of joints in fibre-reinforced plastics: II—Adhesively bonded joints, *Composites,* **13** (1), 29–37 (January 1982).
8. HART-SMITH, L. J., Further developments in the design and analysis of adhesive-bonded structural joints, *Joining of Composite Materials: Proceedings of the Symposium,* Minneapolis, ASTM publication STP 749, April 16, 1980.
9. BOWDITCH, M. R. and STANNARD, K. J., Adhesive bonding of GRP. Symposium on jointing in fibre-reinforced plastics, *Composites,* **13** (3), 298 (July 1982).

10. Parker, B. M. and Waghorne, R. M., Surface pretreatment of carbon fibre-reinforced composites for adhesive bonding. Symposium on jointing in fibre-reinforced plastics, *Composites,* **13** (3), 280 (July 1982).
11. Lubin, G., *Handbook of Composites,* Van Nostrand Reinhold Company, New York, 1982.
12. Crane, L. W., Hamermesh, C. L. and Maus, L., Surface treatment of cured epoxy graphite composites to improve adhesive bonding, *SAMPE Journal,* **12,** 6–9 (Mar–Apr 1976).
13. Hart-Smith, L. J., Mechanically fastened joints for advanced composites—phenomenological considerations and simple analyses, *Fibrous Composites in Structural Design,* Plenum Press, New York, pp. 543–74, 1980.
14. Godwin, E. W. and Matthews, F. L., A review of the strength of joints in fibre reinforced plastics—Part 1, Mechanical fastened joints, *Composites,* **11,** 155–60 (July 1980).
15. Matthews, F. L. and Leong, W. K., The strength of riveted joints in CFRP, *Proceedings of ICCM3,* Pergamon, Paris, pp. 1241–6, 1980.
16. Chang, Fu-Kuo, Scott, R. A. and Springer, G. S., Strength of mechanically fastened composite joints, *Journal Composite Materials,* **16,** 470–94 (November 1982).
17. Wong, C. M. S. and Matthews, F. L., A finite element analysis of single and two-hole bolted joints in fibre reinforced plastic, *Journal Composite Materials,* **15,** 481–91 (September 1981).
18. Hyer, M. W. and Lightfoot, M. C., Ultimate strength of high load capacity composite bolted joints, *Composite Materials: Testing and Design. Proceedings of the Fifth Conference,* New Orleans, ASTM STP 674, pp. 118–36, March 20–22, 1978.
19. Garbo, S. P. and Ogonowski, J. M., Strength predictions of composite laminates with unloaded fastener holes, *1979 AIAA Conference,* Paper 79–0800, pp. 310–14, 1979.
20. Agarwal, B. L., Static strength prediction of bolted joint in composite materials, *1979 AIAA Conference,* Paper 79–0799, pp. 303–9, 1979.
21. Collings, T. A., On the bearing strengths of CFRP laminates, *Composites,* **13** (3), 241–52 (July 1982).
22. Collings, T. A. and Beauchamp, M. J., Bearing deflection behaviour of a loaded hole in CFRP, *Composites,* **15** (1), 33–8 (January 1984).
23. Bashford, D. P., *Guidelines on basic aspects of jointing technology for advanced fibre reinforced plastics used in space structures,* Fulmer Research Laboratories Report R878/15A/October 1984 prepared under European Space Agency Contract 5497/83/NL/PB(SC).

Chapter 8

RHEOLOGY OF CARBON FIBRE COMPOSITE PREPREG MATERIALS

TSUNEO HIRAI

Department of Mechanical Engineering, Doshisha University, Kyoto, Japan

SUMMARY

Carbon fibre composite prepreg materials are essentially heterogeneous structures and moulding them results in unstable deformation. The flow state during a press forming process can be analysed by representing the material flow as being unsteady.

For an unstable condition, a Lagrangean description should be applied, but to simplify the analysis a progressive step-by-step formulation method is developed using an Eulerian description.

By minimising energy, the eigenvalue and linear incremental analyses can be applied to predict the progress of flow for the initial condition. The numerical results reveal the initiation of the weld line which can lead to fracture of the composite structure. An additional outcome of the work is the development of a hybrid moulding method, which avoids this trouble by employing laminates composed of materials of different stiffness.

The progressive flow of a composite during moulding has been analysed numerically by assuming the flow to be anisotropic. The numerical results obtained here by the progressive step-by-step method are shown to be in good agreement with experiments. The rheological method should also be effective for dealing with carbon fibre composite prepreg materials.

8.1. INTRODUCTION

Carbon fibre prepreg materials have been in use since the late 1940's, but in the future they should achieve still greater prominence in the form of sheet moulding compounds because of their suitability for mass production processes. The key to successful fabrication is in applying computer aided design to analyse the rheology of the flow state during the moulding process and so avoid any problems which could jeopardise the large investment.

Unlike homogeneous materials, carbon fibre prepreg composite materials have a heterogeneous structure with distinct interfaces. These may be classified as microscopic surfaces within a lamina, and macroscopic surfaces between laminae. The latter have no penetrating reinforcing fibres across the interfaces. An accurate evaluation of the mechanical behaviour of the interfaces should enable the flow properties of composites to be better understood. Knowledge of the interfaces is particularly important, since flow resistance at a surface results in a pronounced increase in fibre orientation parallel to the surface, and a decrease in orientation in the perpendicular direction. In moulding fibre reinforced materials particular attention should be paid to the variations of fibre orientation caused by the flow state, with the aim of avoiding failure and warp, that is, out-of-plane distortion of the product. The direction of rotation of fibres depends on the geometry of the flow channel, with an opposite tendency according to whether the flow is convergent or divergent as represented by the extrusion ratio. The former condition occurs in press moulding and the latter in injection moulding. This study is concerned with press moulding and the convergent stage of transfer moulding, where the extrusion ratio* is greater than unity.

8.2. PRINCIPLES OF THE MOULDING PROCESS AND ASSOCIATED PROBLEMS

The principal purpose of the study is to develop a design procedure for the moulding process.

For the designer, a precisely controlled ply charge of prepreg produces a complex one-piece structure with well defined mechanical

* Extrusion ratio: the ratio of the cross-sectional area of charged material A_0 to that of the extruded product A. $R = A_0/A$.

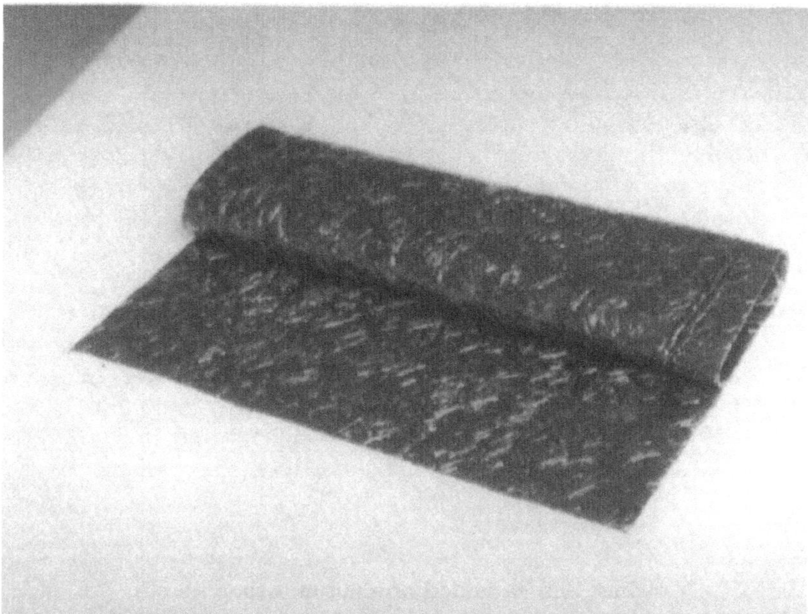

FIG. 8.1. Example of tape-type carbon fibre prepreg material.

properties and an optimised design can be developed with confidence. For the fabricator, prepregs provide an easy-to-handle dry material with a short cycle time, ideal for mass production. Faults in products can be avoided by a knowledge of the rheology of the moulding process.

The newly developed tape-type carbon fibre prepreg, in the form of sheet moulding compound, is shown in Fig. 8.1 and an example of a product is given in Fig. 8.2.

This component was produced by a transfer moulding process using the method illustrated in Fig. 8.3(a), with economical use of material, which is very expensive. Many components can be produced by press moulding between matched dies, as shown in Fig. 8.3(b), to achieve reproducible accuracy and quality for complex shapes in mass production quantities, without the need for skilled operators.

When composite materials are used in compression moulding processes problems can arise due to heterogeneity, caused by fibre orientation during flow. In practice the following faults might be observed

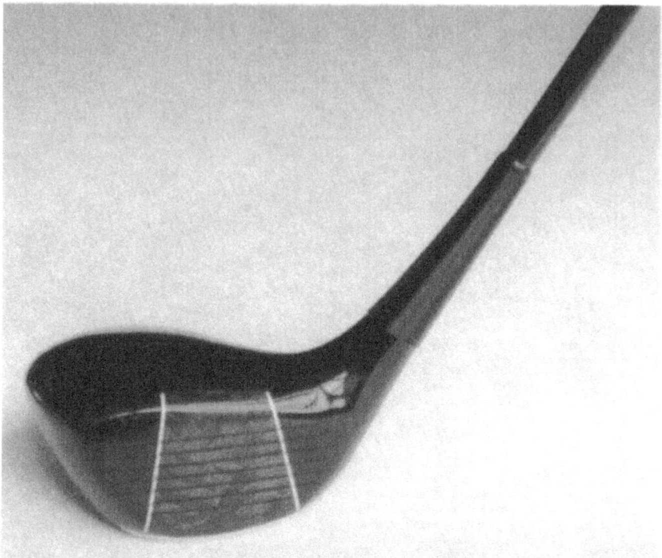

FIG. 8.2. Moulded product in carbon fibre.

in moulded products:

(i) Surface sink, where surface cavities are produced on convex faces during the final stages of moulding.
(ii) Brittle regions, which occur at the ends of fins due to stress concentrations in the resin-rich parts.[1–3]
(iii) Cracks may be initiated at weld lines, which are created in regions between converging flows where the fibre direction becomes normal to the surface.[4,5]
(iv) Warp, that is, out-of-plane displacement by residual stresses, might be caused by an imbalance in temperature distribution in the product at the final stage of the process.

To overcome these problems, by improvements in mould design and forming conditions, it is necessary to investigate the flow characteristics of composite materials and devise means of analysing flow, temperature distribution, etc. These aspects have been investigated analytically, using finite element formulations.

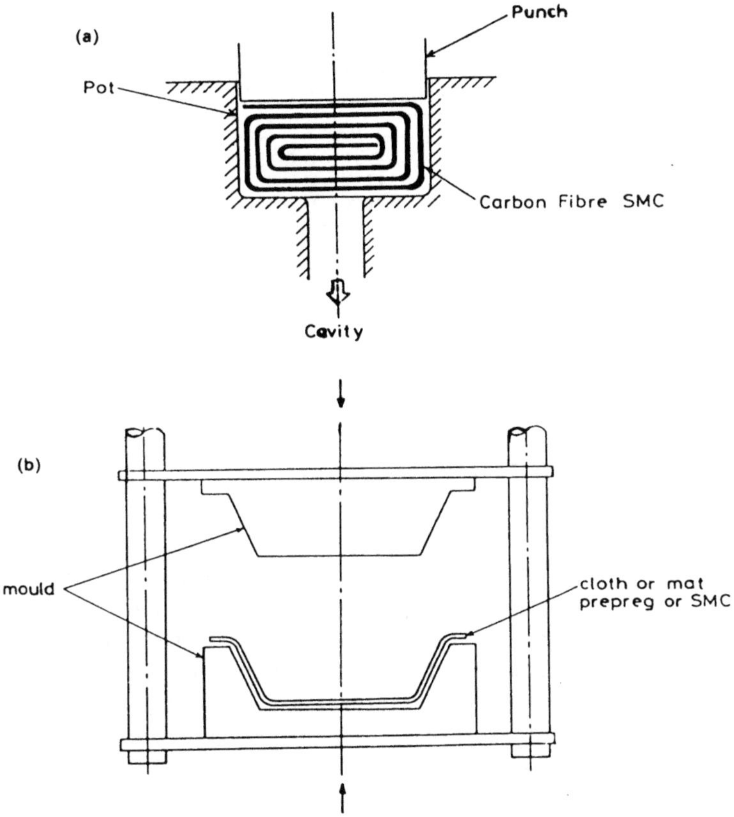

FIG. 8.3. Illustrating transfer and press moulding for carbon fibre prepreg materials. (a) Transfer moulding; (b) heating and pressing in the press moulding machine.

8.3. RHEOLOGY OF COMPOSITE PREPREG MATERIALS

8.3.1. Governing Equations and Formulation

The moulding process applied to composites is essentially unsteady and it is necessary to take account of the orientation of the reinforcement during forming. It will be shown how modelling can be applied to provide a relatively simple method of analysing and designing the process for orthotropic media. The progressive step by

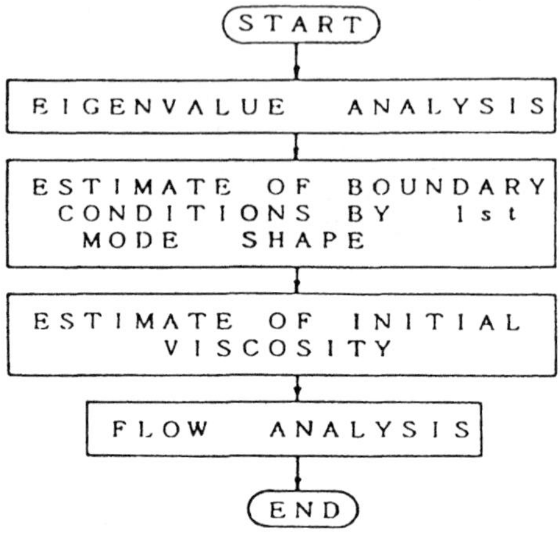

FIG. 8.4. Flow chart for numerical procedure.

step formulation method is employed to simulate the mechanical behaviour of the composite during the unsteady moulding process. The modelling of the mechanical behaviour during deformation is achieved as follows (see Fig. 8.4). For the first step, the deformation state does not depend on the compression rate of the punch, but is governed by the eigenmode of deformation. Thus an estimate of the deformation state before the initiation of flow is calculated by eigenvalue analysis.[6]

$$[K][\psi] = [M][\psi][\Omega^2] \tag{1}$$

$$[\psi] = [\psi_j] \qquad j = 1, 2 \cdots n \tag{2}$$

$$[\Omega^2] = \mathrm{diag}\,[\omega_j^2] \qquad j = 1, 2 \cdots n \tag{3}$$

$[K]$ is the stiffness matrix and $[M]$ the mass matrix. Also $[\Omega^2]$, $[\omega_j]$ and $[\psi]$ are eigenvalue, natural frequency and mode matrix, respectively. The substitution of unit matrix in the mass matrix yields a standard eigenvalue problem from the generalised one. For the numerical procedure on the problem, the sub-space method is applied.

The first mode should be used to investigate the subsequent process, because the energy level of the first mode is lower than the next. The result is then applied to the following step. The sequence of deformation is then simulated by linear incremental analysis, as an infinitesimal deformation due to punch pressure.

In the linear analysis the vertical displacement at the punch surface is given by the punch movement and the horizontal displacement is estimated by the first mode of eigenvalue analysis. From the result the distortion of the finite elements and the relative displacement between punch and material are obtained. The strain rate is determined and the initial viscosity can be estimated. The analysis for the flow process is then applied, as outlined below.

The flow state of composite materials cannot be expressed by the type of flow equation used for a metal or a resin, since they are not isotropic materials but include fibrous reinforcement. Therefore, on the assumption that with composite materials the fibre orientation yields different flow resistance in directions normal and parallel to the flow, the homogeneous orthotropic pseudo-plastic equations are derived and used as governing equations to analyse the flow state. It is also assumed that a fibre orientates along a stream line when the extrusion ratio is greater than unity.

The pseudo-plastic viscosities K_L and K_T in directions parallel and normal to flow, are transformed into K_x and K_y by the transformation matrix,

$$\begin{Bmatrix} K_x \\ K_y \end{Bmatrix} = \begin{bmatrix} \cos^2\theta & \sin^2\theta \\ \sin^2\theta & \cos^2\theta \end{bmatrix} \begin{Bmatrix} K_L \\ K_T \end{Bmatrix} \tag{4}$$

where θ is the fibre orientation with respect to the x co-ordinate.

As the moulding process is unsteady, the progressive step by step method should be used at each stage. According to the progress (e.g. an increment of punch stroke), finite elements are added in the flow direction. The anisotropic pseudo-plastic and continuity equations are written as follows, by assuming incompressibility and absence of body force

$$\rho u \frac{\partial u}{\partial x} + \rho v \frac{\partial u}{\partial y} - \frac{\partial \tau_{xx}}{\partial x} - \frac{\partial \tau_{xy}}{\partial y} = 0 \tag{5}$$

$$\rho u \frac{\partial v}{\partial x} + \rho v \frac{\partial v}{\partial y} - \frac{\partial \tau_{yx}}{\partial x} - \frac{\partial \tau_{yy}}{\partial y} = 0 \tag{6}$$

$$\frac{\partial u}{\partial x} + \frac{\partial v}{\partial y} = 0 \tag{7}$$

where u_j (u and v), τ_{ij} and ρ are the velocity components, the deviator of stress tensor and the density, respectively.

Here, it is assumed that the flow of the composite material is non-Newtonian

$$\sigma_{eq} = K(\dot{\varepsilon}_{eq})^n = K(\dot{\varepsilon}_{eq})^{n-1}\dot{\varepsilon}_{eq} \tag{8}$$

where σ_{eq} and $\dot{\varepsilon}_{eq}$ are uniaxial equivalent stress and strain rate, respectively, and K is the material constant.[7] The relationship between stress and strain rate in the case of pseudo-plastic flow can also be written[8,9]

$$\sigma_{eq} = 3\mu(J_2)\dot{\varepsilon}_{eq} \tag{9}$$

where

$$\dot{\varepsilon}_{eq} = \sqrt{4J_2/3}, \qquad 4J_2 = 2\left(\frac{\partial u}{\partial x}\right)^2 + 2\left(\frac{\partial v}{\partial y}\right)^2 + \left(\frac{\partial u}{\partial y} + \frac{\partial v}{\partial x}\right)^2$$

and J_2 is the invariant of the strain rate tensor.

By combining with eqn (8)

$$3\mu(J_2) = K(\dot{\varepsilon}_{eq})^{n-1} = K(\sqrt{4J_2/3})^{n-1} \tag{10}$$

Then the coefficient of viscosity is obtained

$$\mu(J_2) = \frac{K}{3}\left(\sqrt{\frac{4J_2}{3}}\right)^{n-1} \tag{11}$$

The pseudo-plastic viscosity K is obtained from the fibre orientation using the transformation matrix eqn (4).

$$\mu_x = \frac{K_x}{3}\left(\sqrt{\frac{4J_2}{3}}\right)^{n-1} \tag{12}$$

$$\mu_y = \frac{K_y}{3}\left(\sqrt{\frac{4J_2}{3}}\right)^{n-1} \tag{13}$$

So, the constitutive equations for the orthotropic pseudo-plastic fluid flow can be written

$$\tau_{xx} = 2\mu_x\frac{\partial u}{\partial x} - p \tag{14}$$

$$\tau_{xy} = \tau_{yx} = \mu_y\frac{\partial v}{\partial x} + \mu_x\frac{\partial u}{\partial y} \tag{15}$$

$$\tau_{yy} = 2\mu_y\frac{\partial v}{\partial y} - p \tag{16}$$

where p is hydrostatic stress or pressure. Consequently the orthotropic pseudo-plastic equations can be written.

$$\rho u \frac{\partial u}{\partial x} + \rho v \frac{\partial u}{\partial y} + \frac{\partial p}{\partial x} - \mu_x(J_2)\left(\frac{\partial^2 u}{\partial x^2} + \frac{\partial^2 u}{\partial y^2}\right) = 0 \tag{17}$$

$$\rho u \frac{\partial v}{\partial x} + \rho v \frac{\partial v}{\partial y} + \frac{\partial p}{\partial y} - \mu_y(J_2)\left(\frac{\partial^2 v}{\partial x^2} + \frac{\partial^2 v}{\partial y^2}\right) = 0 \tag{18}$$

$$\frac{\partial u}{\partial x} + \frac{\partial v}{\partial y} = 0 \tag{19}$$

The method of weighted residuals (a Galerkin formulation) is used to solve the equations. The numerical process is shown in Fig. 8.4.

For analysing the flow state with die pre-heating and shortening of the holding time after moulding, in order to reduce the cycle time, the temperature distribution in the material should be described. For this purpose, the following basic energy equation is applied. The viscosity of the material decreases according to the temperature and a numerical analysis is performed over the temperature range until the polymerisation reaction occurs, as shown in Fig. 8.5.

$$\rho c\left(\frac{\partial T}{\partial t} + u\frac{\partial T}{\partial x} + v\frac{\partial T}{\partial y}\right) - \frac{\partial}{\partial x}\left(k_x \frac{\partial T}{\partial x}\right) - \frac{\partial}{\partial y}\left(k_y \frac{\partial T}{\partial y}\right) = 0 \tag{20}$$

T is the temperature, k_j is the thermal conductivity and c is the specific heat. Using the temperature distribution obtained by eqn (20) the dependence of the viscosity coefficient on temperature is defined as follows[10]

$$\mu(\dot{\varepsilon}, T) = \mu(\dot{\varepsilon}) \,.\, \mu(T) \tag{21}$$

Since $\mu(T)$ can be written as

$$\mu(T) = \exp\left[\beta(T_0 - T)\right] \tag{22}$$

where T_0 is the room temperature and β is the material constant, $\mu(\dot{\varepsilon}, T)$ is given by

$$\mu(\dot{\varepsilon}, T) = \mu(\dot{\varepsilon}) \exp\left[\beta(T_0 - T)\right] \tag{23}$$

The temperature distribution is calculated for the initial state, and the viscosity is determined from this. The next velocity distribution step is calculated using the derived viscosity distribution. These procedures are repeated by the progressive step by step method, as shown in Fig. 8.6 for corresponding time increments, with alternate application of

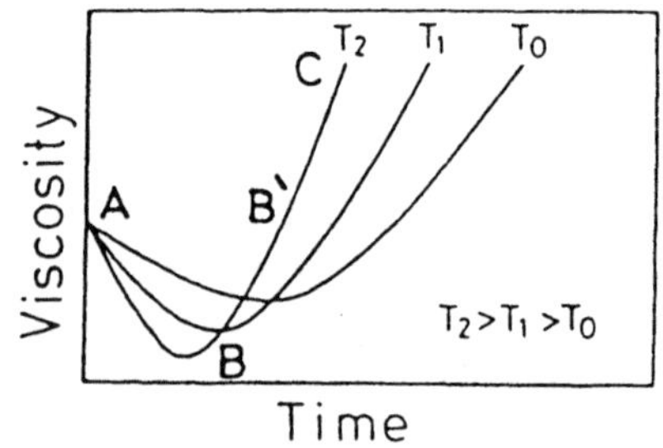

FIG. 8.5. Viscosity–time relationships for material.

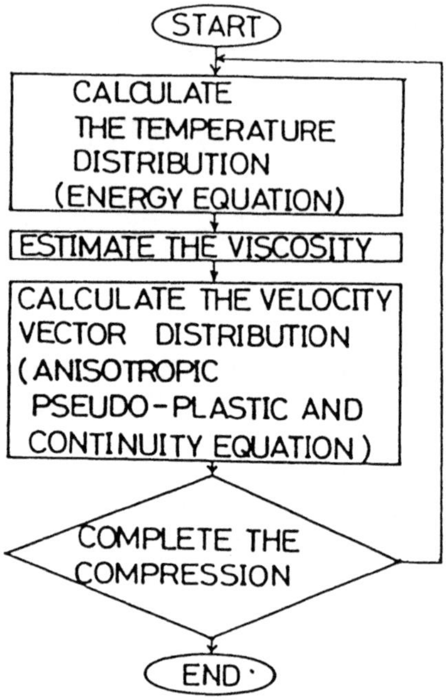

FIG. 8.6. Flow chart for incorporating temperature effect.

the thermal diffusion analysis and the flow analysis in the same interval of incremental time.

8.3.2. Finite Element Formulations

The method of weighted residuals of Galerkin and the Newton–Raphson method are used to solve the non-linear partial differential eqns (17), (18) and (19), as it seems to be extremely difficult to solve them by exact analytical methods.[11–13] Here the equations are rearranged to give eqn (24)

$$\begin{aligned} f_1(u, v, p) &= 0 \\ f_2(u, v, p) &= 0 \\ f_3(u, v, p) &= 0 \end{aligned} \tag{24}$$

The Newton–Raphson method is applied as follows

$$\begin{Bmatrix} \dfrac{\partial f_1}{\partial u} & \dfrac{\partial f_1}{\partial v} & \dfrac{\partial f_1}{\partial p} \\ \dfrac{\partial f_2}{\partial u} & \dfrac{\partial f_2}{\partial v} & \dfrac{\partial f_2}{\partial p} \\ \dfrac{\partial f_3}{\partial u} & \dfrac{\partial f_3}{\partial v} & \dfrac{\partial f_3}{\partial p} \end{Bmatrix}^{(n)} \begin{Bmatrix} u^{(n+1)} - u^{(n)} \\ v^{(n+1)} - v^{(n)} \\ p^{(n+1)} - p^{(n)} \end{Bmatrix} + \begin{Bmatrix} f_1 \\ f_2 \\ f_3 \end{Bmatrix}^{(n)} = 0 \qquad (25)$$

where (n) is the iterative step number and the matrix is Jacobian. Therefore, the formulation obtained by applying the method of weighted residuals is

$$\int_e [W]^T\{[J]^{(n)}\{\Delta\phi\} + \{f\}^{(n)}\}\,\mathrm{d}V = \{0\} \qquad (26)$$

where

$$\{\phi\} = \{u, v, p\}^T, \qquad \{\Delta\phi\} = \{\phi\}^{(n+1)} - \{\phi\}^{(n)}$$
$$\{f\}^{(n)} = \{f_1^{(n)}, f_2^{(n)}, f_3^{(n)}\}^T$$

and $[J]^{(n)}$ is the Jacobian matrix, $[W]$ is the shape function or weighting function, and $\int_e \mathrm{d}V$ is the integration in an element.

Function $\{\phi\}$ is denoted as follows; when shape functions of u, v, and p are represented by M, N and L respectively

$$\{\phi\}_e = \begin{bmatrix} M_i & 0 & 0 & M_j & 0 & 0 & M_k & 0 & 0 & \cdots \\ 0 & N_i & 0 & 0 & N_j & 0 & 0 & N_k & 0 & \cdots \\ 0 & 0 & L_i & 0 & 0 & L_j & 0 & 0 & L_k & \cdots \end{bmatrix} \{\Phi\}_e \qquad (27)$$

where $i, j \cdots k$ are nodal numbers and $\{\Phi\}_e$ are nodal quantities. $[k]^{(n)}$ and $\{f\}^{(n)}$ are denoted, after applying Green's theorem, as follows

$$[K]_e^{(n)} = \int_e [W]^T[J]^{(n)}[W]\,\mathrm{d}V \qquad (28)$$

$$\{F\}_e^{(n)} = \int_e [W]^T\{f\}^{(n)}\,\mathrm{d}V \qquad (29)$$

By the Galerkin method, eqn (26) is represented as

$$\int_e [W]^T\{[J]^{(n)}[W]\{\Delta\Phi\}_e + \{f\}_e^{(n)}\}\,\mathrm{d}V = \{0\} \qquad (30)$$

Therefore, eqn (30) is superposed on the whole domain

$$[K]^{(n)}\{\Delta\Phi\} = \{F\}^{(n)} \tag{31}$$

where

$$\{\Phi\}^{(n+1)} = \{\Phi\}^{(n)} + \{\Delta\Phi\} \tag{32}$$

Initially, $\{\Delta\Phi\}$ is obtained as a linear solution using eqn (31) and the calculation is repeated until $\{\Phi\}$ converges sufficiently. Since it is known that the relation between velocity direction and pseudo-plastic viscosity is non-linear, due to the assumption of the fibre orientation rotating along the streamline, the calculation using eqn (31) is carried out by the repeated substitution method until θ in eqn (4) and the velocity direction converge in value.

The calculation formulation is adopted in alternated order rather than the normal sequence to simplify the treatment for non-linear partial differential equations.

At first the Newton–Raphson method is applied to linearise these equations and the modified equations describing the correction as the unknown function are obtained. The method of weight residuals of Galerkin is applied to discretise the modified equations as given by eqn (31).

8.3.3. Verification of Calculation System

To improve mould design and forming conditions it is necessary to investigate the flow characteristics of composite materials during unsteady forming processes. The practical case of forming a T-shaped component from a flat blank, in a closed die, has been considered as a simple example. Problems encountered in practice include central plane weakness, thermal gradients in the formed component at the ejection stage and non-symmetrical flow. These aspects have been investigated analytically, using finite element formulations, and experimentally using glass reinforced plastics instead of carbon fibre prepreg material and plasticine, which is a homogeneous material, for comparison purposes. It is too difficult to measure the displacement of reference measuring points during the moulding of carbon fibre composites because of its black colour. The microscopically different appearance of carbon fibre and glass fibre is illustrated in Fig. 8.7.

The differences are in their cross section configuration and their surface condition, carbon fibre being treated by sizing and glass fibre covered by coupling agents, in their original forms before curing. The

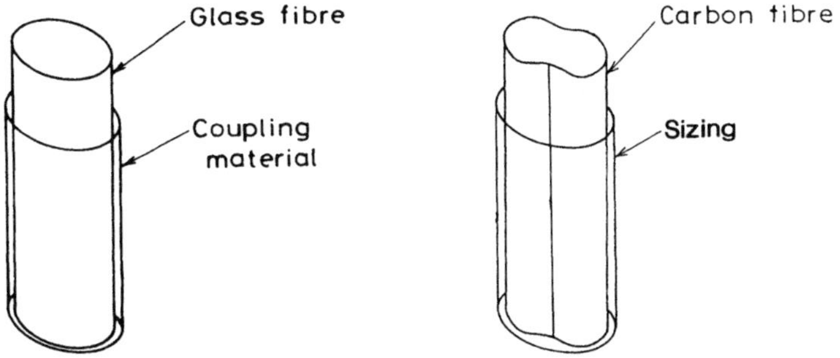

FIG. 8.7. Cross sectional configurations of carbon fibre and glass fibre and their surface treatments.

mechanical properties of the structure have varying characteristics according to the different types of surface treatment. Cracks or fractures due to stress concentrations, arising through the close proximity of fibres in the bundle, occur and develop at the microscopic interfaces between fibre and matrix under load. Therefore the mechanical properties of composites could be influenced at the microscopic interfaces by surface treatment of the fibres.

Since the reinforcement behaves as fibre bundles during flow it is more important in flow analysis to consider the flow resistance at the boundaries of the fibre bundles, i.e. at the macroscopic interfaces.

Therefore the viscosity of the macroscopic interface controls the flow characteristics of prepreg materials. The anisotropic stiffness also affects the initial deformation during the first stage of the moulding process.

If verification of the numerical system has been carried out on glass reinforced plastics the mechanical behaviour of the carbon fibre composite can be represented by suitably adjusting the anisotropic stiffness and the anisotropic flow resistance caused by fibre orientation. The effect of anisotropic characteristics and die geometry on deformation behaviour during moulding will be discussed in the next section.

Since development of a weld line in the early stage of a moulding process has a most harmful effect on a component, it is desirable that the cause is established, and verified by the flow analysis developed in this chapter. Cracks which originate and progress across vertical plane sections of composites with ribs appear to be caused by the stress concentrations produced at the weld lines. The weakness at the weld line occurs because the interface has no interconnecting fibre rein-

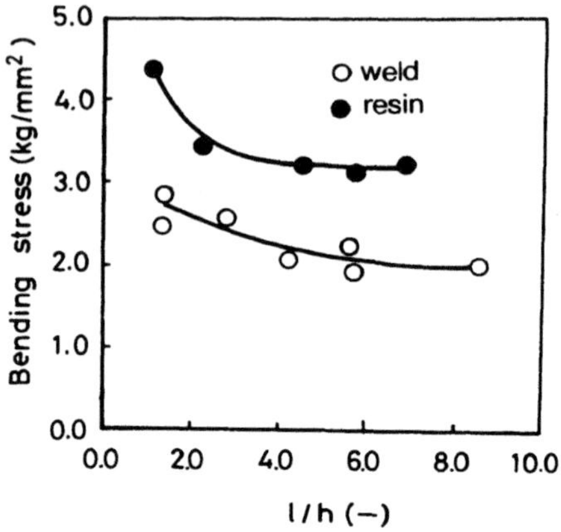

FIG. 8.8. Relationship between bending stress and bending strain.

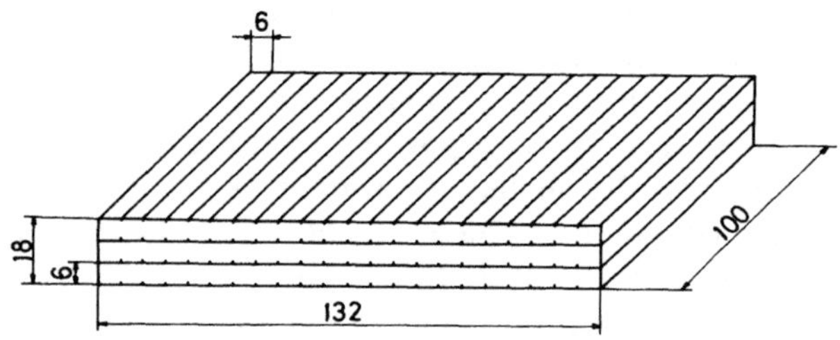

FIG. 8.9. Details of specimen, showing positions of markers.

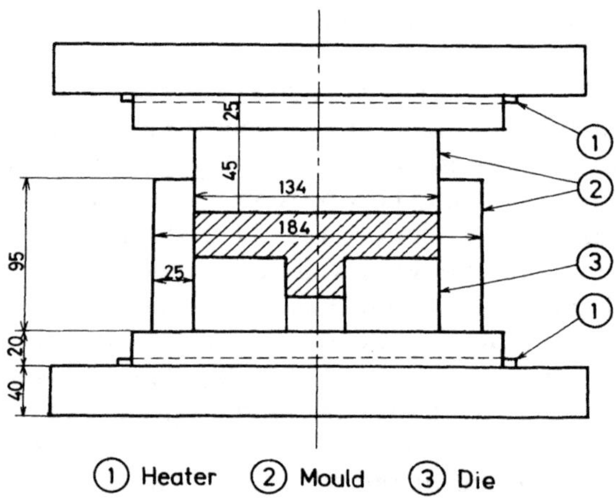

FIG. 8.10. Die set for experiments (forward extrusion).

forcement, being formed by the folding of a surface. Results of bend tests on flat specimens, produced by cutting the ribs from T-shaped components with weld lines, are compared with the results from resin specimens without reinforcement in Fig. 8.8. The reinforced specimens show a decrease in strength rather than an increase, because of the stress concentration caused by the weld line. It is necessary to understand the mechanism of initiation of the weld line so as to avoid this trouble.

To obtain clearly marked measuring points for observing displacements glass fibre prepreg materials of different colours are used for each lamina. Sheet moulding compound (SMC) and bulk moulding compound (BMC), prepressed into similar sheet form, are laminated after black markers have been set in parallel lines on the surfaces, as shown in Fig. 8.9. A schematic drawing of the die set for the experiments is shown in Fig. 8.10. Sectional views of formed components are shown in Fig. 8.11.

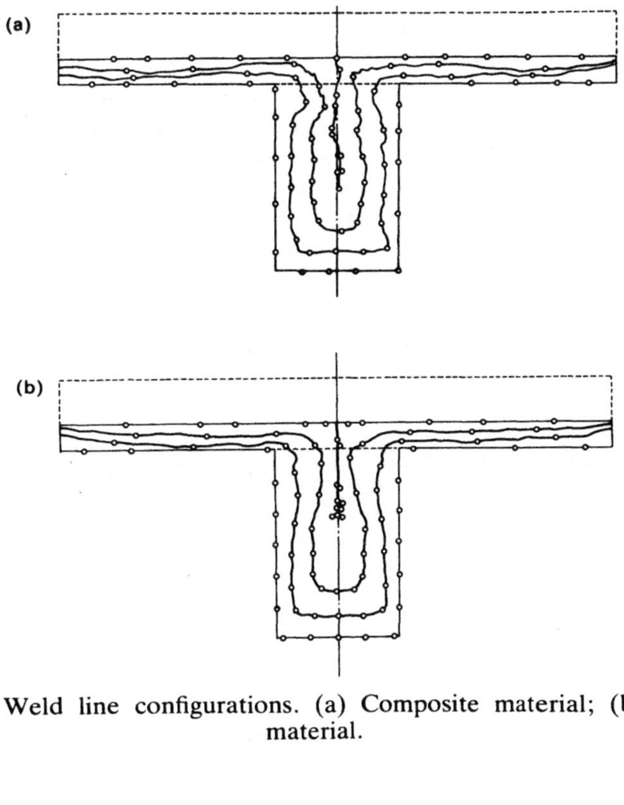

FIG. 8.11. Weld line configurations. (a) Composite material; (b) isotropic material.

It is considered that the meandering form of the central line about the symmetrical axis, as shown in Fig. 8.11(a), is due to the heterogeneity of the composite material, because it does not occur with homogeneous isotropic materials such as metals. The appearance for an isotropic material is shown in Fig. 8.11(b).

8.3.4. Numerical Results for Constant Temperature Moulding

This chapter is concerned with the development of a numerical procedure for analysing the phenomenon, as shown in Figs 8.4 and 8.6. The finite element division of the part, which is a T-shaped flow channel, is shown in Fig. 8.12. A–B–C and D–E–F are surfaces of the die, A–F is the punch surface and C–D is a free boundary over the rib cavity. It is assumed that the composite has orthogonal anisotropic properties dependent on the orientation of the reinforcement. The modulus in the horizontal direction is assumed in this case, from experience, to be three times that in the vertical direction, as the reinforcement is orientated in the horizontal direction. Figure 8.13 shows the first mode obtained by the eigenvalue analysis. This indicates that the nodal points shift not only in a vertical direction, but also horizontally, immediately on loading. Secondly, the linear incremental deformation theory is used to calculate the progressive deformation state. The dots shown in Fig. 8.14 indicate where the reaction forces at nodal points on the punch surface change from compression to tension. The phenomenon indicates the possibility of micro-buckling at the surface. The asymmetrical appearance of dotted nodal points confirms the tendency for snaking of the weld line about the symmetrical axis. Moving on from the initiation of flow, the second invariant J_2 of the strain rate tensor is calculated from the strain rate, which is obtained from the derived strain following the incremental

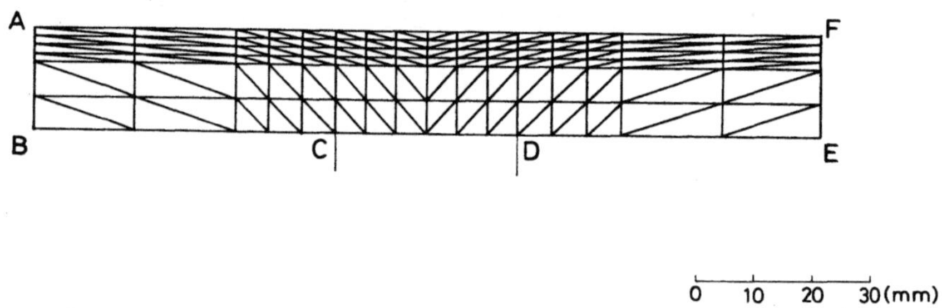

FIG. 8.12. Finite element division.

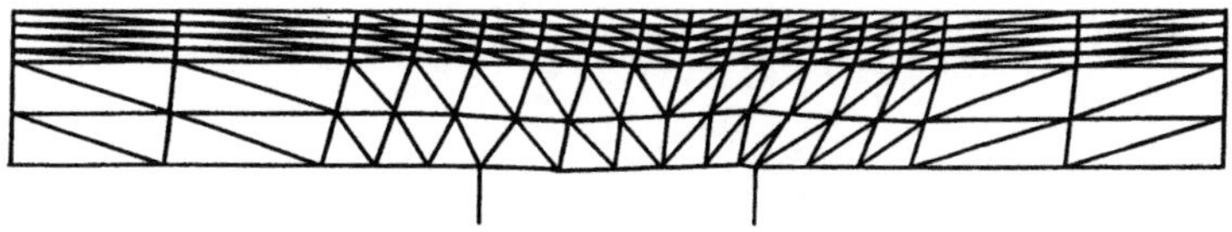

FIG. 8.13. Mode shape obtained by eigenvalue analysis.

analysis. In order to determine the flow characteristics of the composite, the ratio of the pseudo-plastic viscosities K_T/K_L is assumed as 3, since this case indicates an anisotropic flow resistance. The deviation of the anisotropic axis is estimated from the gradient of the element, determined by the above analysis, as shown in Fig. 8.14. Then the initial viscosity for the flow analysis can be calculated. The pattern of

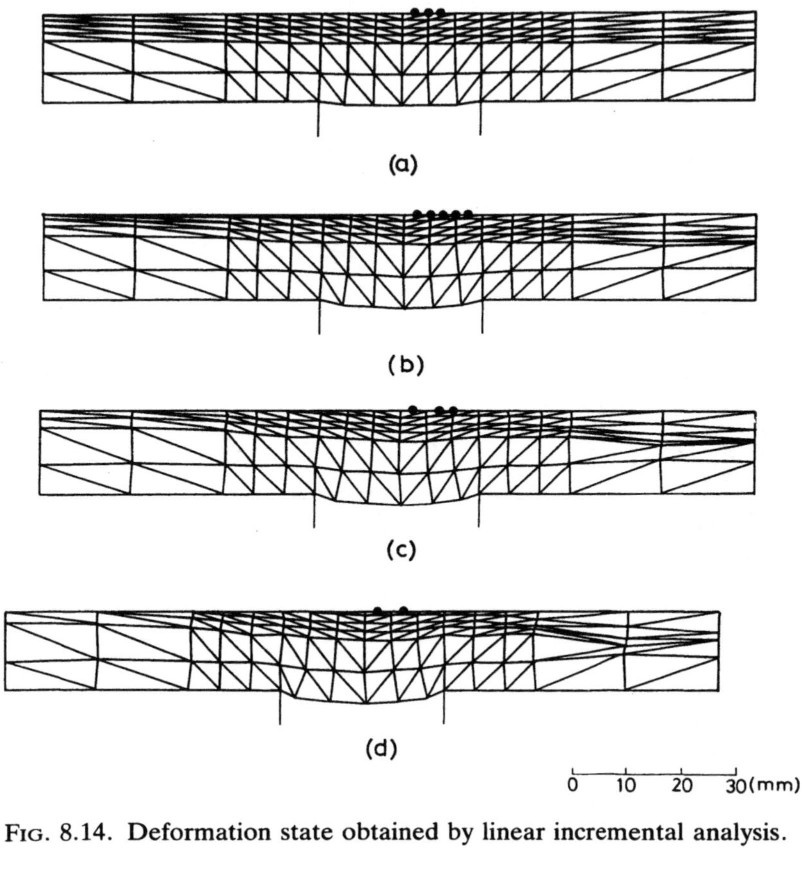

FIG. 8.14. Deformation state obtained by linear incremental analysis.

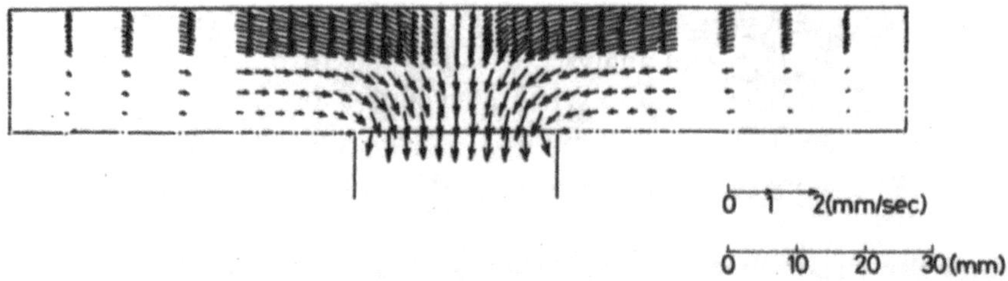

FIG. 8.15. Initial velocity vectors for initial viscosity obtained from incremental analysis.

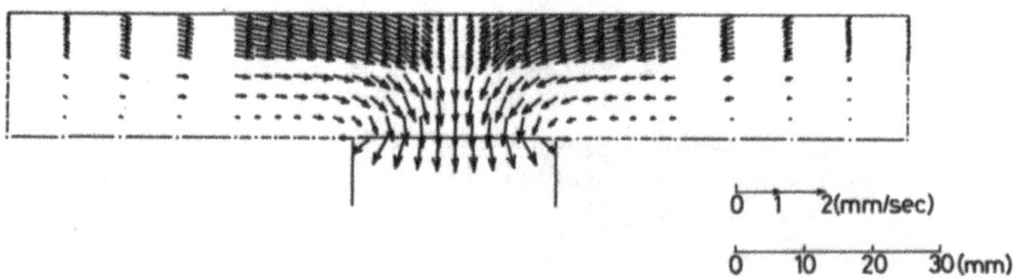

FIG. 8.16. Initial velocity vectors for uniform initial viscosity.

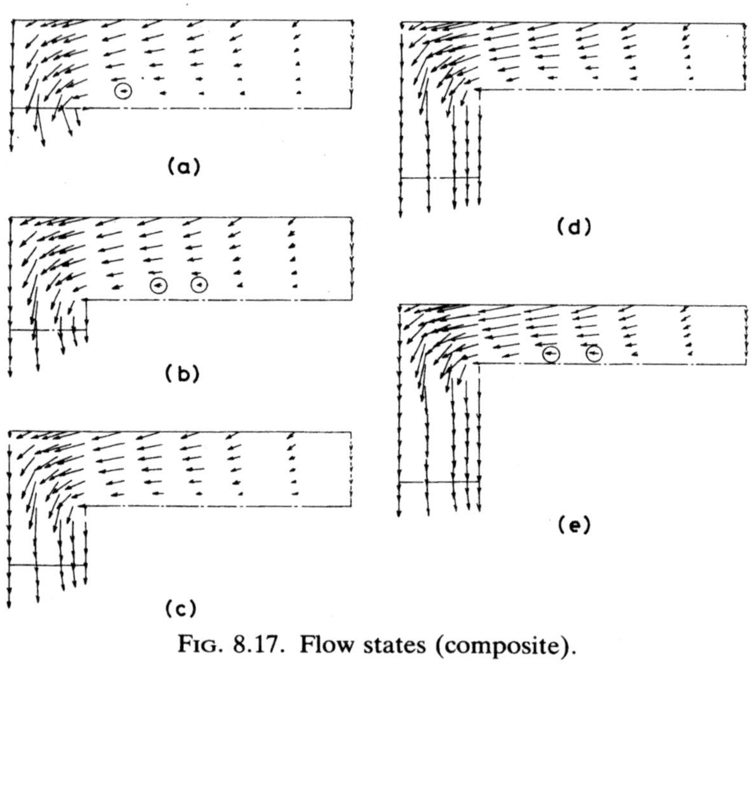

FIG. 8.17. Flow states (composite).

initial velocity vectors, including the effect of initial viscosity obtained by linear incremental analysis, is shown in Fig. 8.15. The flow pattern is asymmetric about the axis and suggests a tendency for disturbance of the flow in successive steps, because the channel in which the material flows is symmetrical. Initial velocity vectors which neglect the microscopic analysis, and assume uniform initial viscosity, result in symmetrical flow, as shown in Fig. 8.16. Numerical results for the flow states during moulding at constant temperature are shown in Fig. 8.17, as an example of unsteady flow using the Progressive Step by Step Method. From Fig. 8.17(b) it is evident that the flow swings around the rib corner, and has a tendency to incline upwards over the horizontal die face. The flow pattern shown in Fig. 8.17(e) indicates a slight meander through the corner into the rib, also a velocity vector with an upward inclination. The flow state suggests that there is a resin-rich region in the product shown in Fig. 8.17. The analytical methods are seen to give good correlation with experiments, shown in Fig. 8.20. Figure 8.18 shows an isotropic flow state, the material constants used being for plasticine, with pseudo-plastic viscosity

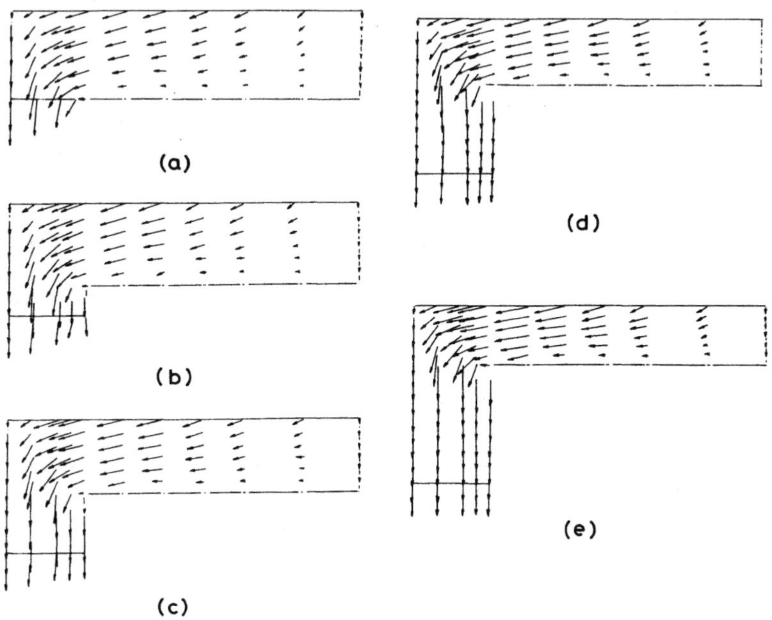

FIG. 8.18. Flow states (isotropic material).

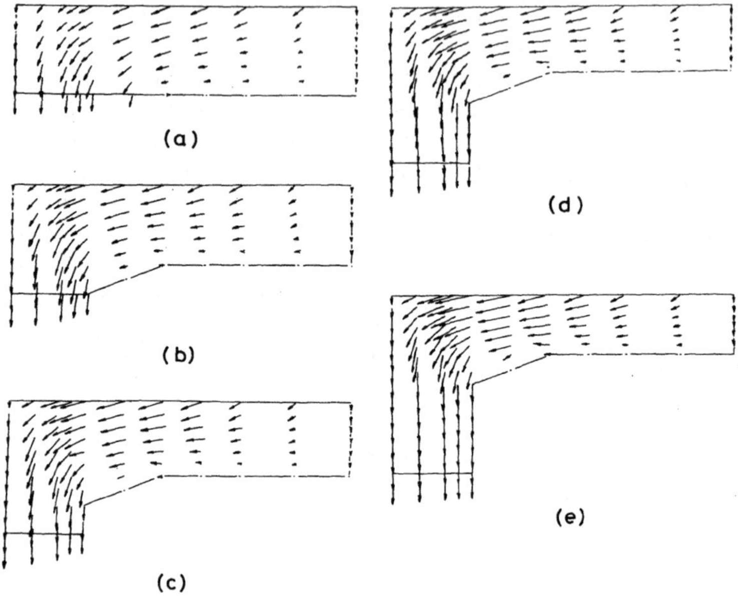

FIG. 8.19. Flow states with corner fillet radius.

$K_T = K_L$. The flow is comparatively smooth, and proceeds around the entrance into the rib channel to the axis of symmetry, showing good correlation between theory and experiment.

In addition, to investigate the effect of a corner radius on the die, a triangular element is incorporated to form a corner fillet, giving the results shown in Fig. 8.19. Comparison with Fig. 8.17 reveals that flow is smoother around the rib corner and meandering flow in the rib is lessened. The horizontal velocity components near the central surface seem to decrease during the final stage of the moulding process. The sinking appears to be caused by this phenomenon, and the tendency becomes less with stronger anisotropy.

8.3.5. Experimental Results for Constant Temperature Moulding

A schematic arrangement of the die set used for experiments is shown in Fig. 8.10. Tests were performed using corner radii of 0 and 6 mm at entry to the rib.

As shown below, the flow state of prepreg or SMC should be analysed with respect to the macroscopic ply interface, since the reinforcement is aligned in the longitudinal direction. Plasticine and

BMC are used as experimental materials for convenience of analysis. The flow state of BMC, as the composite material, is compared with that of the plasticine, which is known to show a behaviour similar to an isotropic metal or the matrix of a prepreg.

The glass fibres of BMC are randomly orientated in the initial state, and become aligned in the flow direction as pressing proceeds. Thus the fibre orientation of BMC is unsteady during the process. In order to prepare specimens with fibres aligned initially in the flow direction, similar to prepreg materials or SMC, the BMC is prepressed from 20 mm thickness to 6 mm. The flow then shows anisotropic characteristics throughout moulding. Experimental data obtained from materials prepressed and laminated as shown in Fig. 8.9 are compared with numerical results.

In order to characterise the flow properties of BMC and plasticine,

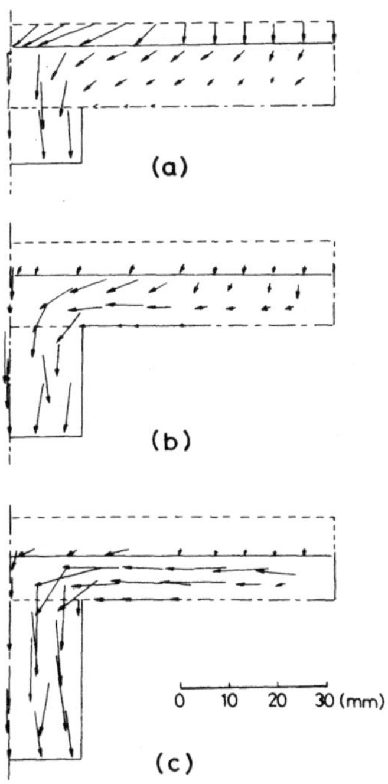

FIG. 8.20. Finite displacement vectors for composite flow and $R_f = 0$ mm (experimental).

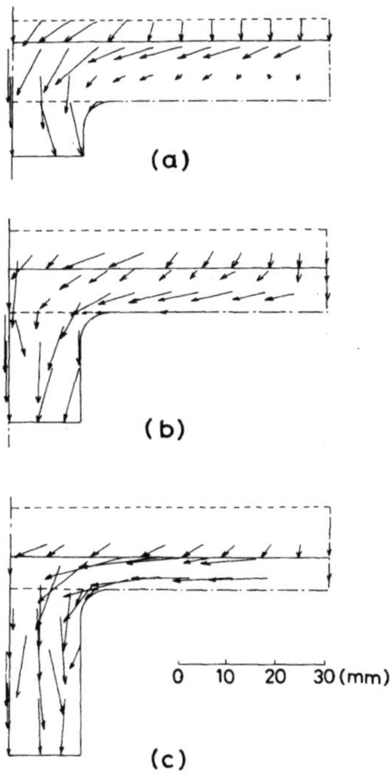

FIG. 8.21. Finite displacement vectors for composite flow using die with corner radius of $R_f = 6$ mm (experimental).

finite displacement vectors can be constructed by observing the locations of markers before and during moulding, as described above. The displacements were determined at three intermediate stages of pressing, designated (a), (b) and (c), at which the length of the rib was 12 mm, 24 mm and 36 mm. For these tests, the press stroke velocity was 0·3 mm/s.

The experimentally determined finite displacement vectors are shown in Figs 8.20 and 8.21 for the composite material, using die corner radii of $R_f = 0$ and 6 mm. Corresponding results for the isotropic material are given in Figs 8.22 and 8.23. From these results it is evident that the flow of the isotropic material is comparatively smooth along the channel and tends to move towards the centre of the rib, whereas the composite material meanders round the corner into

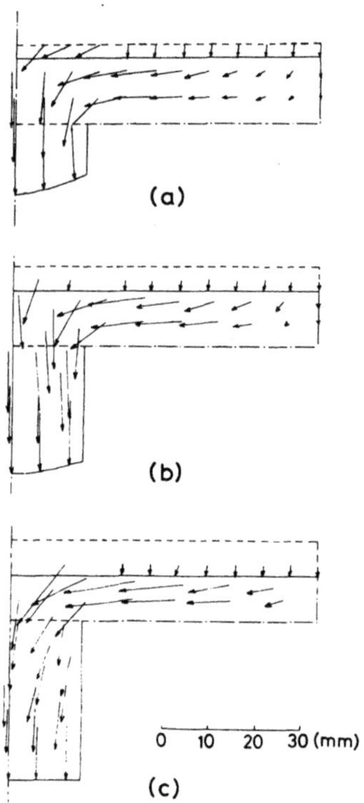

FIG. 8.22. Finite displacement vectors for isotropic material flow (plasticine) and $R_f = 0$ mm (experimental).

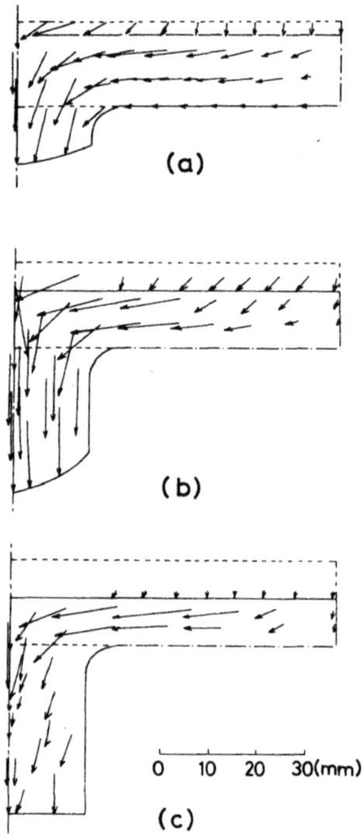

FIG. 8.23. Finite displacement vectors for isotropic material flow (plasticine) using die with corner radius of $R_f = 6$ mm (experimental).

the rib. It is also evident that with increasing corner radius R_f these tendencies are slightly enhanced.

The validity of the analytical model used for representing flow of a composite and involving anisotropic stiffness and anisotropic pseudo-plastic flow behaviour, has been examined by comparing with experimental results. On this basis, the flow state of carbon fibre prepreg materials and the design of the moulding system might be evaluated.

In the initial flow state, the trends indicated by the eigenvalue analysis and the linear incremental and flow analyses are in good agreement with the experimental results, and also with practical experience. For anisotropic composite flow, the initiation of micro-buckling and development of a snaking weld line is observed, and the flow has a tendency to meander through the rib channel. For isotropic material the flow is smooth and tends to move to the centre of the rib channel.

8.3.6. Influence of Thermal Diffusion

In a commercial operation the die is pre-heated and this creates temperature variations in the material which influence the flow behaviour. The aim is to optimise the temperature distribution in the die in order to minimise the moulding cycle time.

As the temperature rises, the viscosity of the material decreases, and then at a certain point it begins to increase because of polymerisation. Thus the analysis of the flow state must be made on the basis of the temperature distribution in the material, arising from heat transfer from the mould. The initial viscosity, for application in the flow analysis will be calculated according to Section 8.3.4 and should then be corrected for temperature changes resulting from thermal diffusion from the heated die.

For the investigations, the mould temperature was 80°C, at which rapid polymerisation would not occur.

An example of the influence of thermal diffusion on the flow state is indicated in Fig. 8.24 and Fig. 8.25, for temperature distribution and corresponding flow state, respectively, with alternate application of thermal diffusion analysis and flow analysis as described above. The tendency for fibres to move slightly upward during horizontal displacement, as shown by the ringed vectors in Fig. 8.17, is reduced under the new conditions, shown in Fig. 8.25.

The numerical results for the models with pre-heated dies are shown to be in good agreement with experimental results shown in Fig. 8.26;

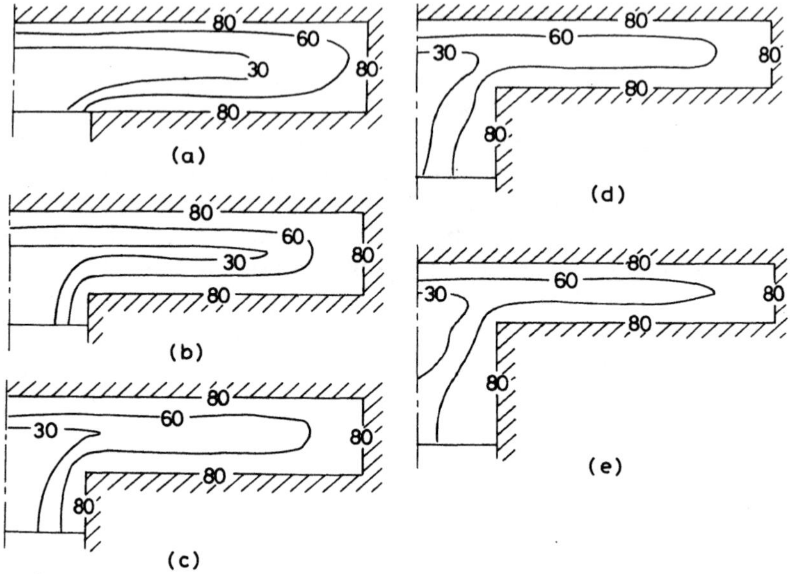

FIG. 8.24. Temperature distribution in the material calculated by the progressive step by step method, for a die temperature of 80°C.

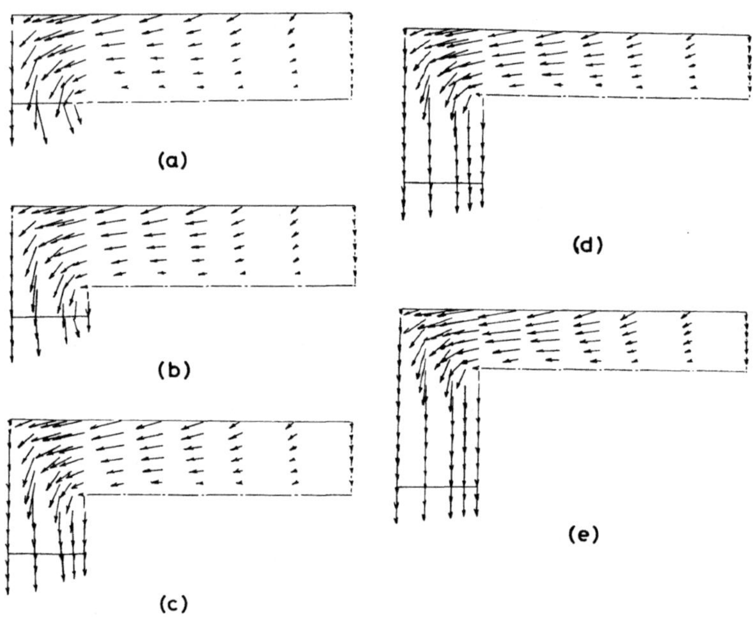

FIG. 8.25. Diagrams of velocity vectors dependent on the thermal states shown in Fig. 8.24.

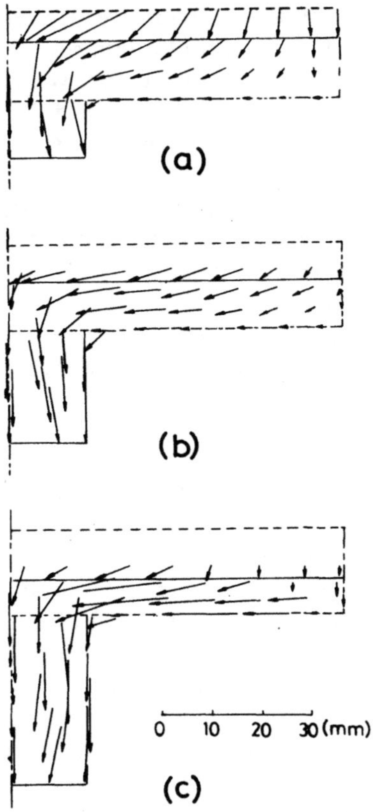

FIG. 8.26. Finite displacement vectors for composite subject to thermal diffusion (experimental).

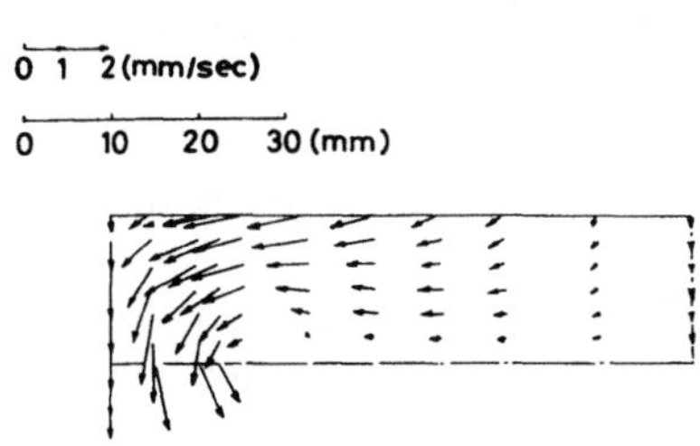

FIG. 8.27. Velocity vectors calculated considering the effect of moulding temperature ($\beta = 0{\cdot}030$).

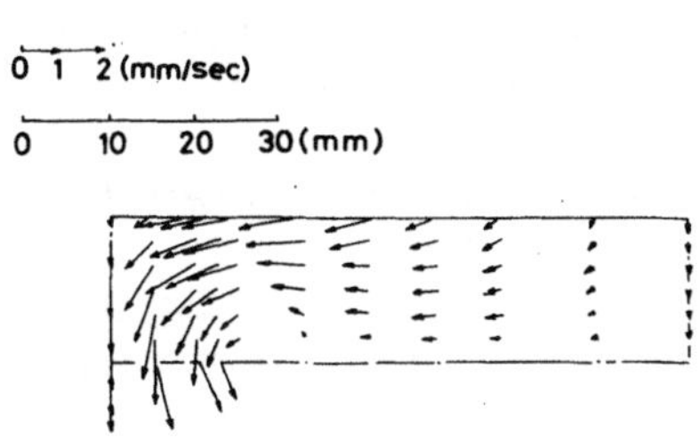

FIG. 8.28. Velocity vectors calculated considering the effect of moulding temperature ($\beta = 0{\cdot}035$).

the surface lamina has a tendency to move towards the symmetric axis and large vectors appear in the middle lamina.

The effect of the value of β on the velocity vectors is shown in Figs 8.27 and 8.28, for the values of $\beta = 0{\cdot}030$ and $0{\cdot}035$, respectively. The decrease in viscosity with increase in temperature is much greater than that found for the previous calculated results where $\beta = 0{\cdot}020$. The flow states of the surface and middle laminae shown in these figures are quite similar to those in Fig. 8.26(a), but the upward flow at the rib entrance is more pronounced. These results indicate that a decrease in viscosity may have the same effect as a pronounced anisotropy. Moreover, regarding the effect of temperature distribution, the greater the temperature difference between the surface and the centre, the more anisotropic the flow state becomes.

8.4. PREDICTION OF ASYMMETRIC FLOW PATTERN

8.4.1. Anisotropy Parameters

The rheology of composite prepreg materials has been developed to enable the flow state to be represented throughout forming and the results have been verified by experiments using glass fibre prepreg materials. The use of anisotropic stiffness and anisotropic flow resistance of the material due to fibre orientation forms the basis of the analysis. The former factor is used to give the initial deformation state by eigenvalue analysis, and subsequent linear incremental analysis, and the latter factor governs the subsequent plastic flow condition.

To apply these two anisotropic characteristics to carbon fibre composite prepreg materials it is necessary to determine the effects of variations of anisotropic stiffness ratio (E_L/E_T) and anisotropic flow resistance[15] (K_T/K_L).

8.4.2. Influence of Anisotropic Stiffness and Flow Resistance on Flow Patterns

The T-shaped die cavity, already described, was used for investigating the influence of anisotropic stiffness and flow resistance on flow patterns.

The first mode shape obtained by eigenvalue analysis, shown in Fig. 8.13, indicates asymmetrical flow, with horizontal displacement of nodal points. The subsequent deformation states obtained by linear

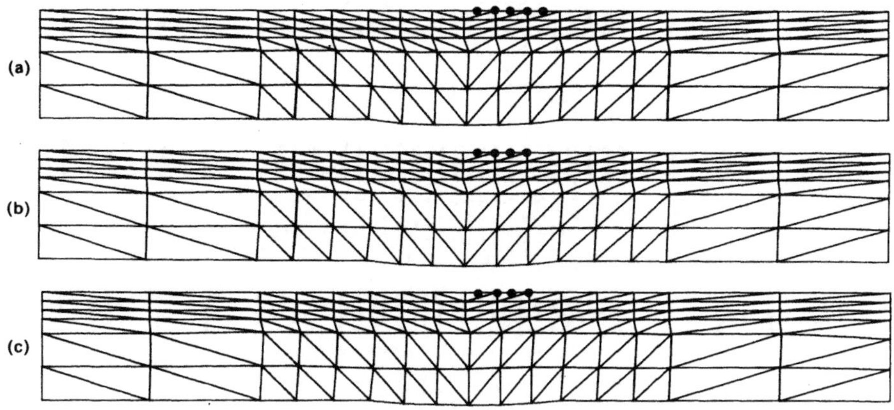

FIG. 8.29. Deformation state obtained by linear incremental analysis for comparison with the result of eigenvalue analysis. (a) $E_L/E_T = 1$; (b) $E_L/E_T = 3$; (c) $E_L/E_T = 5$.

incremental analysis, for different anisotropic stiffness ratios E_L/E_T of 1, 3 and 5, are shown in Fig. 8.29 for a punch displacement of 0·1 mm. The results indicate that the deformation modes are similar for the different ratios. The dots shown in the figure indicate where the reaction force at nodal points on the punch surface change from compression to tension. The occurrence of these pressure-free nodal points at progressive stages of deformation are shown in Fig. 8.30.

For the isotropic case, Fig. 8.30(a), the pressure-free nodal points cover the widest range at the first stage, and then reduce in extent whereas for an anisotropic stiffness ratio of 5 the range of pressure-free nodal points remains almost constant during deformation.

With increase in anisotropy, the extent of the pressure-free surface, and the likelihood of microbuckling at the surface, is seen to be greater, particularly in the later stages of compression. The actual

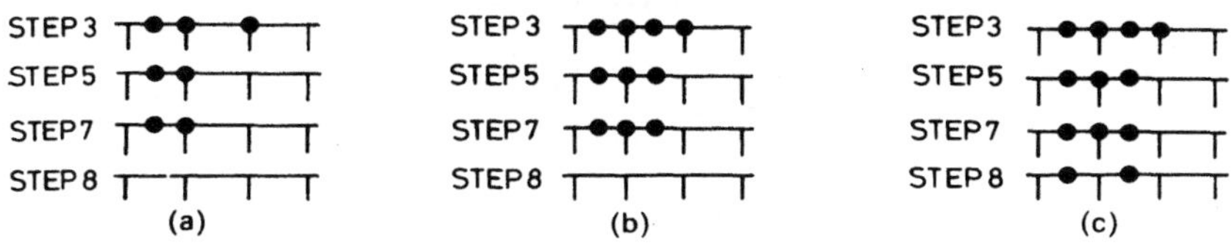

FIG. 8.30. Pressure-free nodes during deformation. (a) $E_L/E_T = 1$ (isotropic); (b) $E_L/E_T = 3$; (c) $E_L/E_T = 5$.

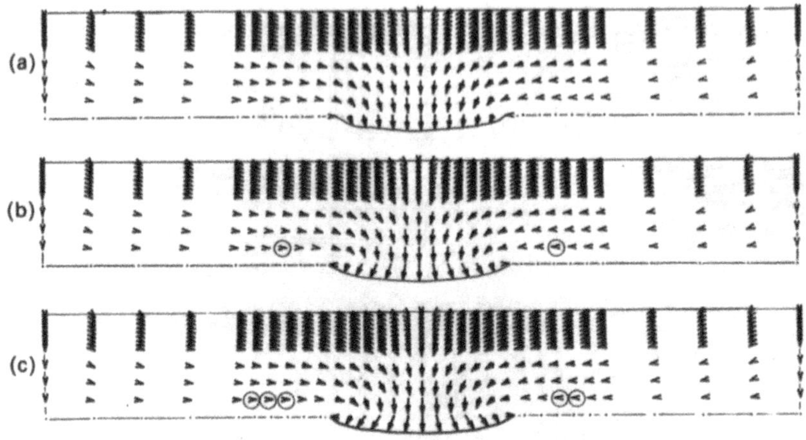

FIG. 8.31. Initial velocity vectors. (a) $K_T/K_L = 1$; (b) $K_T/K_L = 3$; (c) $K_T/K_L = 5$.

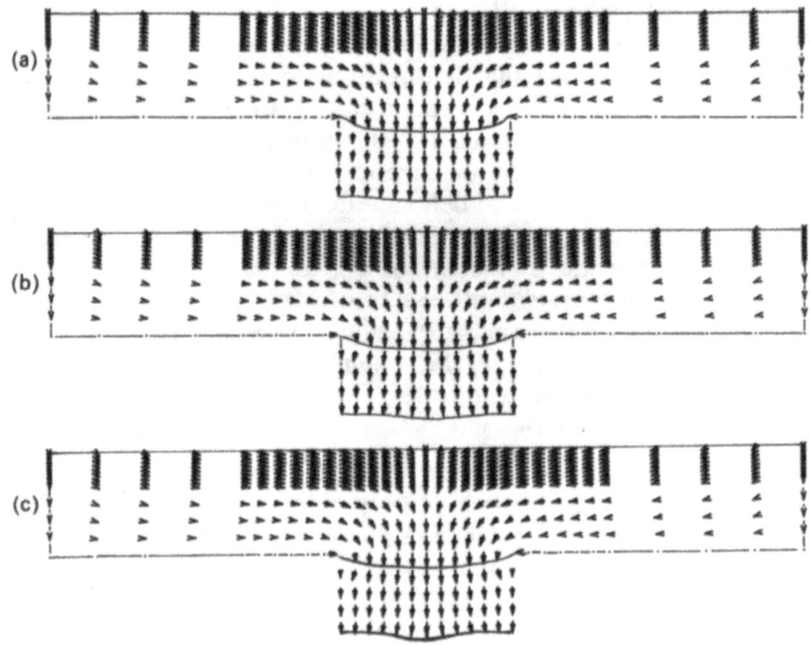

FIG. 8.32. 3rd step velocity vectors. (a) $K_T/K_L = 1$; (b) $K_T/K_L = 3$; (c) $K_T/K_L = 5$.

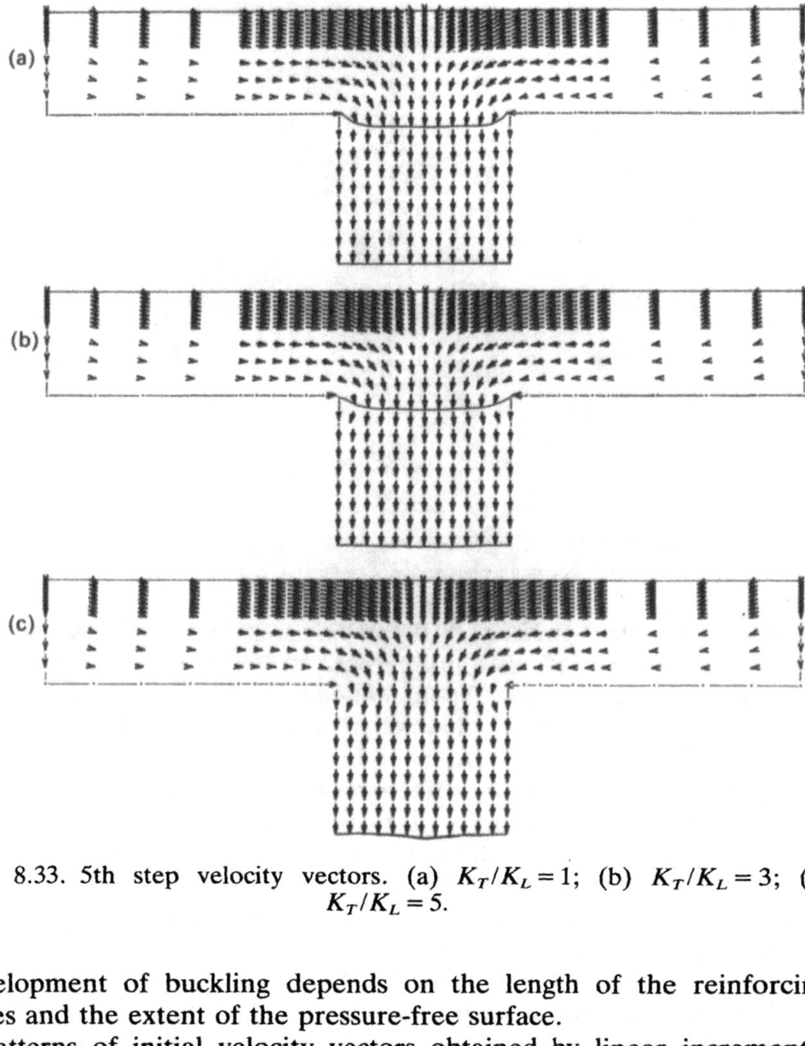

FIG. 8.33. 5th step velocity vectors. (a) $K_T/K_L = 1$; (b) $K_T/K_L = 3$; (c) $K_T/K_L = 5$.

development of buckling depends on the length of the reinforcing fibres and the extent of the pressure-free surface.

Patterns of initial velocity vectors obtained by linear incremental analysis are shown in Fig. 8.31 for values of anisotropic flow resistance K_T/K_L of 1, 3 and 5. The velocity vectors in the rib become inclined outward from the symmetrical axis with increase in anisotropy. Also, material flowing towards the rib, over the horizontal die face has a tendency to incline upward with the greater factor of anisotropy, as shown by the ringed vectors in Fig. 8.31, and the flow swings sharply around the rib corner.

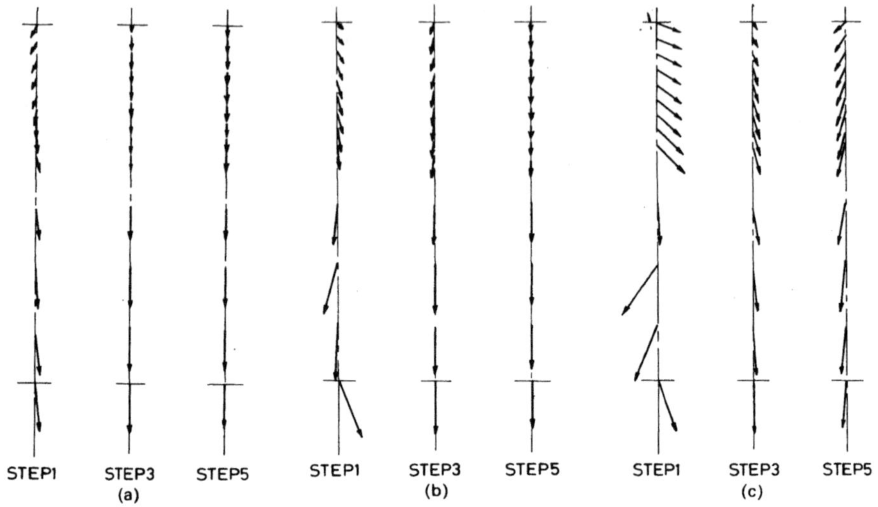

FIG. 8.34. Fluctuation of velocity vectors about centre line of rib. (a) $K_T/K_L = 1$; (b) $K_T/K_L = 3$; (c) $K_T/K_L = 5$.

The velocity vectors at the 3rd and 5th step of the progressive step by step method are indicated in Figs 8.32 and 8.33, respectively. Points of velocity vectors have been linked together to indicate flow distributions across the rib channel. A fairly uniform flow is evident for the isotropic case ($K_T/K_L = 1$) in Fig. 8.32 but for anisotropic conditions the velocities increase towards the centre of the channel. The reason may be that stress transmission in the fluid is less with increase of anisotropy because the principal axis of anisotropy lies parallel to the central axis of the rib.

The distribution becomes uniform after further flow, irrespective of the anisotropy, as shown in Fig. 8.33.

It is well known that the weld line has a tendency to meander and the investigation reveals an increase in the magnitude of the velocity vectors along the symmetric axis, as shown in Fig. 8.34. The swing of the vectors in the upper region of the rib is greater with increase in anisotropy factor, but decreases as the flow proceeds. However, for the case of $K_T/K_L = 5$ the swing persists throughout the process.

8.4.3. Effect of Hybrid Lamination on Flow State

In order to demonstrate the influence on flow of the interlaminar regions where there are no interpenetrating fibres, a study was made of the moulding of a composite consisting of separate layers of isotropic matrix material and anisotropic plies of composite material. An example of the results of the eigenvalue and linear incremental analyses on the deformation state is shown in Fig. 8.35 and the pattern of initial velocity vectors following this condition is given in Fig. 8.36. The flow pattern suggests a greater tendency for asymmetric flow about the axis than the example shown in Fig. 8.31(c).

Similar results for the hybrid lamination illustrated in Fig. 8.37 are shown in Fig. 8.38 and Fig. 8.39. An anisotropic stiffness factor of 3 is assumed, and the stiffness of material A is 3 times that of B. An anisotropic flow resistance factor of $K_T/K_L = 3$ is used. The results suggest a possible means of preventing the development of a weld line in a component.

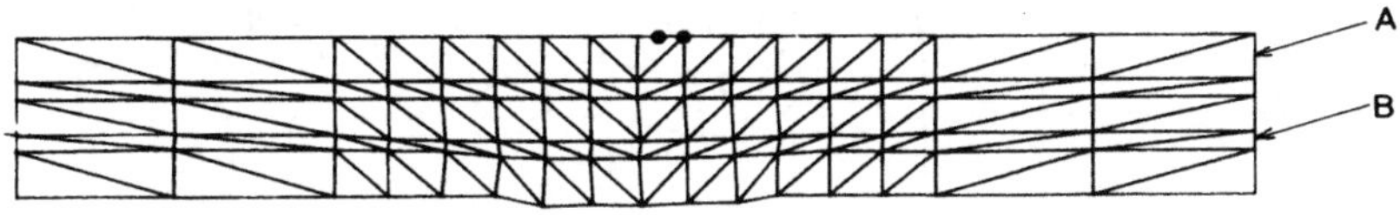

FIG. 8.35. Deformation state obtained by linear incremental analysis for a hybrid model with isotropic interlaminar elements. (Each A is a composite lamina and B a macroscopic resin-rich interface.)

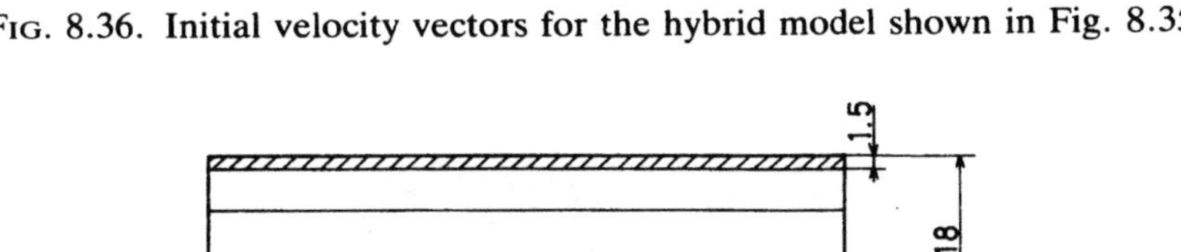

FIG. 8.36. Initial velocity vectors for the hybrid model shown in Fig. 8.35.

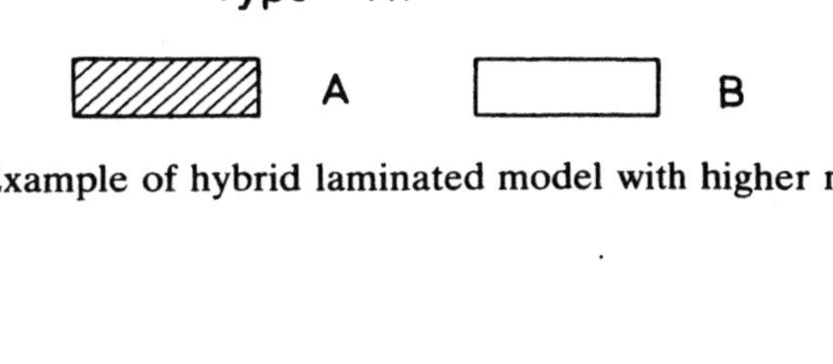

FIG. 8.37. Example of hybrid laminated model with higher modulus at A.

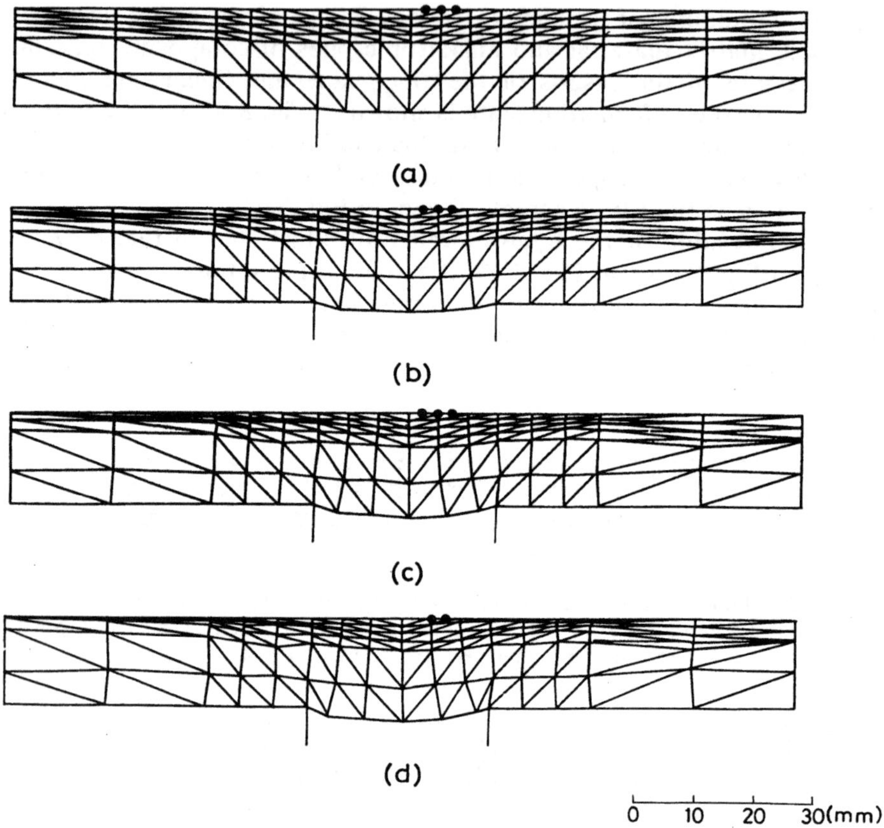

FIG. 8.38. Deformation state obtained by linear incremental analysis.

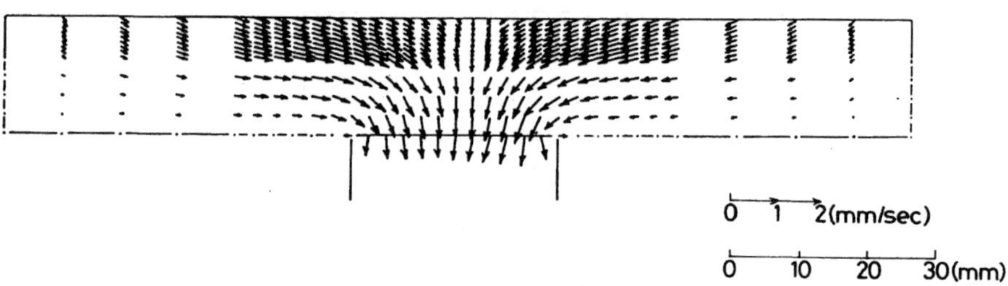

FIG. 8.39. Initial velocity vectors for the hybrid laminated model (Fig. 8.37).

REFERENCES

1. Hirai, T. and Katayama, T., *Proc. of the 1978 Int. Conf. on Composite Materials,* p. 1233, 1978.
2. Hirai, T., Katayama, T. and Hamada, H., *Advances in Composite Materials,* ICCM-III, 1606 (1980).
3. Hirai, T., Katayama, T. and Hamada, H., *Progress in Science and Engineering of Composites,* ICCM-IV, 1593 (1983).
4. Hirai, T., Hamada, H. and Yokoyama, A., *Proc. of ICF Int. Conf. on Fracture Mechanics,* p. 1069, 1983.
5. Quigly, J. J., *Proc. of 37th Annual Conf. SPI,* p. 14-b, 1982.
6. Bathe, K. and Wilson, E. I., *Numerical Methods in Finite Element Analysis,* Prentice-Hall Inc., Englewood Cliffs, NJ, 1976.
7. Burns, R. and Gandhi, K. S., *Proc. of 32nd Annual Technical Conference,* The Society of the Plastic Industry, p. 7-c, 1977.
8. Tomita, T., *Japan Society of Mechanical Engineers Trans. 25,* **149,** 17 (1959).
9. Bird, R. B., *AI Ch E J.,* **5,** 569 (1959).
10. Pervadchuk, V. P. and Yankov, V. I., *Heat Transfer—Soviet Research,* 10.1.11, 1978.
11. Tanaka, T., Hirai, T. and Katayama, T., *Proc. of Second International Symposium on Finite Element Method in Flow Problems,* p. 275, 1976.
12. Hirai, T. and Katayama, T., *The Science and Engineering Review of Doshisha University,* **18** (1), 43 (1977).
13. Hirai, T. and Katayama, T., *The Science and Engineering Review of Doshisha University,* **20** (2), 61 (1979).
14. Myers, F. A. and Maxel, J. M., *Proc. of 33rd Annual Technical Conference.* The Society of the Plastic Industry, p. 16-a, 1978.
15. Hirai, T., Katayama, T. and Hamada, H., *J. of Japan Society for Technology of Plasticity,* **25** (287), 113 (1985).

INDEX

GPSR Compliance

The European Union's (EU) General Product Safety Regulation (GPSR) is a set of rules that requires consumer products to be safe and our obligations to ensure this.

If you have any concerns about our products, you can contact us on ProductSafety@springernature.com

In case Publisher is established outside the EU, the EU authorized representative is:

Springer Nature Customer Service Center GmbH
Europaplatz 3
69115 Heidelberg, Germany

Batch number: 10607212

Printed by Printforce, the Netherlands